# Sustainable Action and Motivation

*Sustainable Action and Motivation* proposes individual competencies and institutional policies that can help overcome the motivational hurdles that hamper sustainable action.

Following the Paris Agreement of 2015 and the UN Sustainable Development Goals, the political momentum urgently to begin the drastic reduction of greenhouse gas emissions has increased significantly. Affluent, high-income OECD countries are expected to take the lead in the global transition to a low carbon society. Given this, we need a better understanding of the motivational problems that people in affluent countries face with acting sustainably. This book investigates the above questions by analysing three fundamentally different perspectives: individuals and their motivation to act sustainably; institutions who take responsibility for issuing policies that steer us towards taking sustainable action; and humanity, each individual member of which ought to understand his or her non-sustainable behaviour in relation to the continued existence of the collective of human beings.

Using theories from empirical psychology and a phenomenological approach to the research, this book will be of great interest to students and scholars of practical philosophy, psychology of motivation and environmental psychology, as well as policymakers looking for ways to implement effective policies that encourage pro-environmental behaviour.

**Roland Mees** studied fundamental mathematics at Utrecht University, business administration at the Rotterdam School of Management (Erasmus University), and philosophy in business at Utrecht University. Alongside his full-time job at ING, he obtained his PhD in 2017 at the Ethics Institute of Utrecht University, where he is an affiliated researcher. Mees is a director of Sustainable Finance at ING, based in Amsterdam, The Netherlands.

## Routledge Environmental Ethics

**Sustainable Action and Motivation**
Pathways for Individuals, Institutions, and Humanity
*Roland Mees*

For more information on the series, please visit www.routledge.com/Routledge-Environmental-Ethics/book-series/ENVE

# Sustainable Action and Motivation

## Pathways for Individuals, Institutions, and Humanity

**Roland Mees**

Routledge
Taylor & Francis Group
LONDON AND NEW YORK

earthscan
from Routledge

First published 2020
by Routledge
2 Park Square, Milton Park, Abingdon, Oxon OX14 4RN

and by Routledge
605 Third Avenue, New York, NY 10017

*Routledge is an imprint of the Taylor & Francis Group, an informa business*



*British Library Cataloguing-in-Publication Data*
A catalogue record for this book is available from the British Library

*Library of Congress Cataloging-in-Publication Data*
Names: Mees, Roland, author.
Title: Sustainable action and motivation : pathways for individuals, institutions and humanity / Roland Mees.
Description: Abingdon, Oxon ; New York, NY : Routledge, 2020. | Includes bibliographical references and index.
Identifiers: LCCN 2019029814 | ISBN 9780367189877 (hardback) | ISBN 9780429199677 (ebook)
Subjects: LCSH: Sustainable development–Psychological aspects–Case studies. | Environmental psychology. | Moral motivation.
Classification: LCC BF353.5.S87 M33 2020 | DDC 155.9–dc23
LC record available at https://lccn.loc.gov/2019029814

ISBN: 978-0-367-18987-7 (hbk)
ISBN: 978-0-429-19967-7 (ebk)

DOI: 10.4324/9780429199677

Typeset in Bembo
by Wearset Ltd, Boldon, Tyne and Wear

# Contents

# Preface

Why would anyone write a book with the purpose of promoting sustainable action when human beings seem incapable of changing their non-sustainable behaviour? Along with Dale Jamieson (2017) and many others, I could have taken a pessimistic stance on the probability that humans will manage to keep global warming below 2 °C, let alone 1.5 °C. I could have carried on living comfortably, well above a minimum threshold of wealth, with my family, job, friends, and musical hobbies, leaving the problem of climate change to people who will live after me. However, I would then not have paid attention to a question to which I increasingly felt that it was up to me to respond.

When I was in the initial phase of my PhD project, I spoke to many colleagues, friends, and family members about questions of intergenerational justice. We discussed whether our moral obligations towards future generations would motivate us to, for example, eat less meat, leave the car at home, or take fewer flights. To my surprise and dismay, I often received an answer along the lines of: "But Roland, what's in it for me?" It seemed that people like me, who are living above a minimum threshold of wealth, would make their commitment to take sustainable action dependent on something they would get in return. I conjectured that, in other words, people in affluent countries who are asked to lower their luxury emissions would mainly do this out of self-interest. In this book, which builds on my dissertation, I investigate as a philosopher what is behind the "What's in it for me?" response. The motivational difficulties of people who want to act sustainably are my focus.

The goal of the book is to give you a glimmer of hope. The motivational problems for individual agents taking sustainable action seem more severe than for people who take another morally complicated action. Moreover, institutional sustainable policies may become easily frustrated due to severe inaction and stalemate among the stakeholders. Yet there are pathways open to us as individuals that could help to strengthen our capabilities to act sustainably. We can make use of scientific research to propose institutional sustainable policies that can be defended by politicians in constitutional democracies. Finally, we can reach a level of self-knowledge in which we integrate our understanding of who we are as a human being with our commitment to make a significant contribution to meeting the challenge of climate change.

At this point, I want to thank those who have supported me during the years of writing this book. First, I am most grateful to my PhD thesis advisors, Joel Anderson and Marcus Düwell. They have supported me during all the stages of the project. I thank Joel and Marcus for sharing their extensive philosophical experience with me and stimulating my development as a scientist.

My gratitude goes out to Dieter Birnbacher and May Thorseth, who, in addition to sitting on my Review Committee, invited me to write a paper for the volume they edited (2015). This paper turned out to be a cornerstone for the project of writing this book. I am grateful to Christoph Baumgartner, Frans Brom, Marc Davidson, Ingrid Robeyns, and Linda Steg, who also sat on my Review Committee. Finally, I am grateful to Rutger Claassen, Boudewijn De Bruin, Liesbeth Feikema, Jesse Mulder, Dirk Schoenmaker, and the members of the colloquium at the Ethics Institute of Utrecht University, who, at various stages of the project, encouraged me and shared their open and candid feedback.

Given that I have completed this book alongside my full-time job at ING, I want to express my gratitude to the bank. Ever since I started with my master's thesis in Philosophy in Business, my colleagues in the Capital & Liquidity Expert Team in Wholesale Banking have always been interested in what I have been doing. More recently, when I moved to Sustainable Finance, my colleagues have supported me in following through with writing this book. All in all, from the years of my philosophy studies until today, ING has given me tremendous opportunities to work on implementing the bank's sustainable strategy by allowing me to practise the pathways that I have investigated in this book.

Finally, I would like to thank Routledge and the series editor for including the book in their Environmental Ethics Series.

# Introduction

> In the middle of the 20th century, we saw our planet from space for the first time. … From space, we see a small and fragile ball dominated not by human activity and edifice but by a pattern of clouds, oceans, greenery and soils. Humanity's inability to fit its doings into that pattern is changing planetary systems, fundamentally. Many such changes are accompanied by life-threatening hazards. This new reality, from which there is no escape, must be recognised – and managed.
>
> (World Commission on Environment and Development 1987, 1)

We start with the first paragraph from *Our Common Future*, the UN report released more than thirty years ago, because, as it seems to me, humanity is still struggling with the "new reality, from which there is no escape". This reality is that global greenhouse gas emissions, mainly caused by that part of humanity living in affluent countries, are still increasing year by year. While investigating some of the problems – "humanity's inability" – my thinking about sustainable action, of which I give an account here, has taught me one thing: it could go the wrong way; our common future could possibly lead to a catastrophe.

In my view, the potential climate catastrophe will not emerge due to insufficient scientific knowledge about climate change; nor will it happen due to a lack of norms that could steer our behaviour. Instead, it may come about because those who accept the outcomes of climate science, and who are in agreement with the principle that we must live up to the duty to save the planet for future generations, nonetheless are stuck in inaction as far as sustainability is concerned. We cannot manage to do what ought to be done, despite the fact that we have agreed that we collectively ought to do it. Therefore, despite the fact that people accept the claims of climate science, and despite the fact that people accept the challenge that calls for human beings collectively to mitigate climate change, the basic question that occupies me is: Why is it hard for people to engage in sustainable action?

This question leads me to investigate the motivation of human agents to take up the challenge of mitigating climate change. Why is it that people seem to have difficulties with taking action, while accepting the norm that we ought to change our non-sustainable behaviour? Why do agents who are usually efficacious in other morally and pragmatically complicated situations seem to be hampered in their motivation to follow through with taking sustainable action? How can institutions effectively implement policies that aim to promote sustainable behaviour if they have insufficient knowledge of the motivational problems with which individuals are faced?

To date and to my knowledge, these questions have been discussed within philosophy by only a handful of philosophers. Given the urgency of the problem that humanity still has not "managed" a sustainable future for the planet, I take it as a sufficient reason for writing this book that we need a better understanding of the motivational problems which people in affluent countries face with acting sustainably. To this end, we will undertake our investigation into the above questions by analysing three fundamentally different perspectives: *individuals* and their motivation to act sustainably; *institutions* who take responsibility for issuing policies that steer us towards taking sustainable action; and *humanity*, each individual member of which ought to understand his or her non-sustainable behaviour in relation to the continued existence of the collective of human beings.

The *first* level of analysis in our investigation is to understand how the motivation of human individuals for sustainable action operates – from the agent's perspective. That is, we are concerned with the interplay, as the agent experiences this, between the collective duty to mitigate climate change, the options available to the agent to take sustainable action, and the agent's motivation to act sustainably. Understanding an agent's motivation from his or her own perspective involves taking the first-person experience of human agents of their own actions in morally complex situations as the basis for the investigation. Within this field, we can study the agent's struggle to succeed with an action that is intended to and does in fact contribute to mitigating climate change (which we will call a sustainable action).

The *second* level of analysis concerns the question how the difficulties with motivation for individuals can be of use to the institutions whose policies are required for promoting sustainable behaviour. Policies that ignore knowledge of the motivational drivers for sustainable action seem to have little chance of being effective, since they might be psychologically unrealistic. Specifically, in constitutional democracies, institutional policymakers must know what drives the motivation for sustainable action from the perspective of individual agents, in order to be effective in putting sustainable policies up for approval by the larger community of these agents. Therefore, on the basis of an understanding of the hindrances from the agent's first-person standpoint, we discuss pathways for institutions that could help agents in moving towards sustainable behaviour. In arguing for such pathways, we relate the findings of our research to existing approaches in philosophy and psychology with regard to institutional sustainable policies.

At a *third* level of analysis, in my view, it is necessary to clarify the implications for how we must understand ourselves as human beings if we persist in non-sustainable behaviour. Suppose that we identify some of the motivational obstacles that agents encounter, and suppose that we implement institutional policies for promoting sustainable action, then according to which conception of who we are as human agents can we understand ourselves if we keep failing to act sustainably? The problems with climate change mitigation that we still face raise important questions about the anthropological consequences of humanity's persistent inaction on this issue. A further question, then, is: How should we each understand our moral integrity as an *individual* if we cannot manage *collectively* to fit our doings into the pattern of the planet due to our failure to follow through with changing our non-sustainable behaviour?

The research topic of this book, namely the motivation of individual agents for taking sustainable action, can be considered from the perspective of practical philosophy as well as empirical psychology. Given that the book remains within the field of practical philosophy, three remarks on methodology are important. First, this investigation is an example of practical philosophy which is empirically informed and which aims to inform the empirical sciences. On the one hand, this involves consulting theories from the *psychology of moral motivation* and *environmental psychology* that deal with the topic of sustainable action. On the other hand, the intention is that formulating the tentative hypothesis in this book will enable further philosophical and empirical research into the question of what drives people when taking sustainable action.

Second, as our investigation is based on the agent's first-person experience of her own actions in morally relevant situations, our main methodological question is: Which methods are available for investigating first-person experience to which practical philosophy could make a contribution? Our goal is to find out which hindrances agents experience when they take action on mitigating climate change while having a perhaps rather weak motive for doing so and while being confronted with contingencies in life that may affect the possibilities they perceive for carrying out their plan. Choosing the agent's first-person perspective as the object of study in this way renders our method *phenomenological*. Different from empirical psychology, which is focused on describing and analysing the behaviour of people, phenomenology analyses the perspectives that people have on the phenomenon of their own experience. For the purpose of this book, we can understand phenomenology, narrowly construed, as an interpretative method that focuses on descriptions of the agent's experiences.

Third, the experiences of agents who take morally complicated actions are conveyed through case study material. We consult philosophical and empirical psychological theories in order to interpret the first-person experiences of agents when performing the morally complicated actions in these case studies. Therefore, the method is *conceptual* in the sense that the analysis clarifies philosophical and psychological concepts by interpreting a real-life human being's agency in the case study. To summarise, interpreting the agent's first-person perspective phenomenologically against the background of concepts from philosophy and psychology means that the method of this book can be considered a *conceptual-phenomenological analysis*.

While investigating the levels of analysis, as roughly described above, by applying the conceptual-phenomenological method, I have formulated the *Core Thesis* of this book as follows:

> ***Core Thesis***
>
> *We are concerned with agents in the affluent countries who have accepted the claim of climate science that the main causes of climate change are anthropogenic; these agents have also accepted that people in the affluent countries are under the moral obligation collectively to substantially reduce their luxury greenhouse gas emissions. This means that our focus is agents in the affluent countries who have accepted what we will call "Humanity's Challenge".*
>
> *While attempting to take up Humanity's Challenge by acting sustainably, individual agents are confronted with motivational problems in each of the action phases, be it during deliberation, when making a commitment, or while executing the action. In order to overcome these motivational problems, individuals ought to take certain measures, supported by institutions that implement sustainable policies in constitutional democratic societies. Finally, if we do not make a structural commitment to taking sustainable action, then we suffer as humanity, and as individual and institutional agents, from severe moral failure.*

This overall thesis will be developed throughout the chapters of the book, which are organised in three parts. Below is an outline of the parts of the book, with a summary of the contents of their respective chapters.

In *Part I*, the foundation for the book is laid. The problem with motivation for sustainable action is elaborated, I discuss the methodological aspects of the investigation, and I present case descriptions of agents taking a morally complicated action, which are used throughout the book. A more detailed overview of Part I runs as follows.

One possible explanation for the limited philosophical research into the problem with motivation for sustainable action may be the amount of groundwork that is required to prepare for a discussion of the problem as such. For example, what should we choose as the concept of sustainability? And, what does it mean to act sustainably? These are just the basic starting questions that must be dealt with. If reasonable positions can be argued for in these debates, then the field of human motivation can be located in relation to the fields of climate science and normative ethics. That is, it is necessary to find out whether there is sufficient ground to assume that the main causes of climate change are indeed anthropogenic, and whether, for example, theories of intergenerational justice can be considered as a normative basis for the task of mitigating climate change. If the answer to one of these questions were to be negative, there would be no reason to discuss the motivational problem. I summarise the discussion of the above questions in a concise statement named Humanity's Challenge, and I label an action 'sustainable' if it actually contributes to meeting Humanity's Challenge and is intended to do so.

Subsequently, we consider the question: Do we have reason to consider the problem of motivation for sustainable action to be a philosophical problem worth investigating, or should it be left to empirical psychology? In general, the question about whether a norm (e.g. Humanity's Challenge) is capable of motivating an agent to act in conformity with it is as old as practical philosophy itself. However, I argue that there are (at least) four relevant features that seem specific to an agent's motivation for sustainable action, which require a philosophical investigation. The first such feature relates to the question: What are the meta-ethical foundations of the problem? I argue that it seems plausible to assume that people currently living have a motive for taking action for the benefit of future people. This motive may, however, be so weak that stronger self-interested motives tend to override it. The question, then, to be worked on in Part II, can be phrased as: What keeps agents in affluent countries from taking the step from accepting Humanity's Challenge to actually taking sustainable action as an *individual* agent? (Chapter 1).

However, before this question can be discussed, we need to respond to the second philosophical feature of the problem with motivation for sustainable action. This concerns the lack of a sharply defined addressee of the moral requirement to take up Humanity's Challenge. In other words, we must be clear about the division of labour between individual agents and institutions. Is it defensible to leave Humanity's Challenge in the hands of institutions only, or ought individuals in affluent countries to accept some responsibility for reducing their luxury emissions irrespective of the institutional policies? In dealing with this question, I take a position on the concept of institutions in that institutions always rely on individual agents for implementing policies. However, this does not imply that the responsibilities of institutions can be reduced to individual responsibilities. Nonetheless, I argue that individual agents bear some responsibility to take sustainable action, which leads to choosing the experience of the individual agent as the methodological starting point of the research (see above).

In order to structure the analysis of the agent's experience of performing a morally complicated action, I have chosen a mainstream descriptive model from the psychology of motivation: the Rubicon model of action phases. This model divides human action into the phases of deliberation, commitment, and execution. The choice of the Rubicon model determines the structure of Part II, in which a chapter is devoted to each of these phases, with an analysis of the motivational features of taking sustainable action that can be assigned to each phase (Chapter 2).

However, before beginning the investigation proper, it is useful to discuss the sort of cases that are relevant to the analysis. I describe five situations of agents who are involved in morally relevant action situations, including sustainable action situations (Chapter 3). Together with the philosophical and psychological theories, the case descriptions form the material to be taken into account while performing the conceptual-phenomenological analysis in the next part.

In *Part II*, we take on the analysis of how the motivation of human individuals for taking sustainable action works as seen from their own perspective. As previously explained, the composition of the chapters follows the phases of the Rubicon model: deliberation, commitment, and execution. Taken together, these chapters provide the descriptive diagnosis of the problem with motivation for sustainable action. The following is a detailed overview of Chapters 4 to 6.

When considering the deliberative phase, note that accepting Humanity's Challenge does not imply that the agent has accepted that she is under the duty as an *individual* to act sustainably. Given this, many things can happen that influence the agent such that she deviates from deciding in favour of a particular sustainable action. This leads to two questions: How can we focus the agent's deliberation by formulating a practical syllogism? And, which cognitive obstacles that hamper people's deliberation about sustainable action are reported in the literature of environmental psychology? Next, we discuss the two remaining philosophical features mentioned in Chapter 1. These include the question whether agents are justified in believing that the norm to take up Humanity's Challenge demands too much from them. The final question considered is: What shapes agents' temptation to delegate their decision about sustainable action to a future self, and thus procrastinate as regards further deliberation? (Chapter 4).

The agent who has passed beyond the phase of deliberation is confronted also in the commitment phase with specific motivational hurdles. Making a commitment means facing the existential question: Am I prepared to do or refrain from doing certain things in order to help mitigate climate change, even though my actions will primarily benefit people who will live in the distant future? In order to clarify the motivational features of acting sustainably, use is made of two philosophical theories in relation to interpreting the case material: Michael Bratman's intrasubjective theory of intention – since the forming of intentions is among the first signs of an agent's commitment – and Axel Honneth's intersubjective theory of recognition – since making commitments leads to changes in one's social relations with other agents. The key questions about intentions are: How strong do the agent's intentions need to be, if we consider their internal consistency and means-end coherence? And, to what extent must the agent believe that she will be effective during the executive phase? The analysis of intersubjective recognition, as developed in the chapter, serves to highlight the complex connections between diminished self-esteem and self-respect and the public expression of one's commitment to sustainable action (Chapter 5).

The last phase of the Rubicon model concerns action execution. When entering the executive phase, there are a variety of ways in which the commitment to carry out a deliberately chosen sustainable action may still run aground. From the intrapersonal perspective, we consider another element of Bratman's theory – the inclination to nonreflectively reconsider an intention – and ask why some agents do so more easily than others. What indications are there that sustainable actions are a special case when it comes to weakness of will and procrastination? Then, we consider theories of self-efficacy and self-control from psychology, which provide insights into the resources that agents need for action execution. As with the discussion of the commitment phase, we end the chapter by discussing an intersubjective aspect of action execution: What would be the effect on one's identity formation if one were confronted with severe resistance from significant other agents while taking a morally relevant action? (Chapter 6).

*Part III* builds on the provisional results of Part II. We conclude the analysis of the perspective of the individual agent by identifying capabilities that help her with overcoming the motivational challenges. The agential capabilities identified for the individual agent constitute the basis for the institutional pathway in the form of policies that promote sustainable action. Finally, while analysing the pathway for humanity, I argue why, despite all the facilitation and effort put in by individuals and institutions, we have to consider the possibility that our will could be morally corrupted when it comes to acting in the interest of future generations. Here is the overview of Chapters 7 to 9.

The descriptive diagnosis that emerged from Part II of the motivational problems for individual agents when taking sustainable action has twofold relevance. First, agents need capabilities for dealing effectively with these hurdles. We identify, therefore, capabilities of practical reasoning and judgement, capabilities of social coordination and recognition, and capabilities of the will, which agents ought to take some responsibility for developing. Second, following the discussion about institutions in Chapter 2, we discuss the institutional support that is needed for the development of these agential capabilities. The institutional strategies we consider are the provision of public information in order to facilitate sustainable action, nudging strategies that steer agents' behaviour while remaining independent from the individual's motivation, and moral education as a means to build the capabilities of agents. However, even when the agential capabilities are in place and the relevant institutional policies implemented, taking sustainable action may remain difficult, specifically in a context with multiple stakeholders. The obstacle that occurs here is severe inaction or stalemate. In response, we formulate a "second-order commitment" that could help in coping with this hurdle (Chapter 7).

On the basis of the proposed agential capabilities that are required for taking sustainable action, and the institutional policies that are meant to support agents, we examine the situation of an agent in an institutional set-up in which changes are needed. We discuss in depth the example of an agent in the banking industry who faces the problem of severe inaction or stalemate because of stakeholder interests drifting apart. We ask: What are the ethical margins of manoeuvre for bankers, when implementing a sustainable policy, given that stakeholders heavily influence them? The case description shows how agents can become caught in a web of ethical and legal stakeholder relationships, which requires them to make a second-order commitment in order to prevent the "on-the-spot burden" of severe inaction. On the one hand, once the second-order commitment is made and the agent receives initial support from the stakeholders, the case study shows the potential for implementing sustainable policies in the banking sector. On the other hand, it shows that the motivational problem can only be taken away in part and that, despite sustainable policies, our ambition may be too modest to contribute significantly to meeting Humanity's Challenge (Chapter 8).

In the final chapter, we take up the question: Could our failure to make a commitment to do something about climate change, which entails our failure to develop the required agential capabilities and institutional policies and strategies, be seen as a form of moral corruption? I discuss Stephen Gardiner's definition of the concept of moral corruption, and propose to reformulate it, inspired by Kant's account of "radical evil". I argue that moral corruption is a self-deceptive form of trying to maintain one's convenient status quo. However, if human beings were to display, en masse, morally corrupt agency, then what would be the effect of such agency on us? I argue that if we lead the life of a morally corrupted agent, even if we are only vaguely aware that we are doing so and consider our lives good, we will pay the price of suffering from a divided practical identity. This kind of agency forms an impediment to our agential capabilities as we exercise them every day, for example in leadership roles in social institutions. In particular, our understanding of moral corruption leads to the question: How should we understand our moral integrity if we fail to take up Humanity's Challenge because we choose morally corrupt agency? If we aim to possess the virtue of integrity, I argue that we have not only the *moral* motive for preserving the planet for future generations, but also the *indirect* motive for making a commitment to take sustainable action. The final question we address is: What kind of commitment to take up Humanity's Challenge do we need to make in order to stay away from morally corrupt agency and remain candidates for possessing the virtue of integrity? (Chapter 9).

With the overview and structure of the book set out, we can now turn to the investigation.

Part I

# Problem, method, and case material

# 1 The problem with motivation for sustainable action

## 1.1 Sustainable action – "What's in it for me?"

Within the range of our daily experiences, there are some that confront us with fundamental decisions about our life. These experiences do something to us; they transform us. They force us to change the view we happen to have on our lives. They make us aware of what we care about. And what we find out unexpectedly is that what we care about may differ from what others, who are significant to us, care about. Here is an example of such an experience.

A group of thirteen homeowners lives in a middle-class town somewhere in an affluent country. The material conditions of their prosperity are very similar. They all have a financial situation sufficiently above a certain threshold of wealth; they can afford to invest €5,000 in the installation on each of their homes of nine solar panels, without having to refrain from any other meaningful investment they might like to consider in order to continue with their lives above that threshold. All the homeowners have chosen to stay in their homes for the longer term. In their pragmatic view, it follows that they have to do something themselves to achieve a certain level of social cohesion, from which they will each benefit, and which is to the benefit of all. Overall, these are responsible, participating citizens in their part of town; their social relations can be characterised as safe, pleasant, and tolerant.

Suppose that this group of homeowners has gathered for a meeting about various topics of common concern and suppose that during this meeting someone raises the topic of climate change. They discuss the question as to whether they ought to install solar panels on their homes, since this would be a meaningful step towards their part of town becoming climate neutral in the long term. All homeowners agree that, from a moral point of view, they have a duty to install solar panels as a contribution to a reduction of the town's greenhouse gas emissions.

Having agreed to this moral claim, S, who is one of the homeowners, calls for action by proposing that they accept an attractive offer from a reputable supplier to install nine solar panels on each house over the next couple of weeks. She explains that although each participant will need to make an upfront payment of roughly €5,000, the payback time is approximately eight years, which is well within the time period the homeowners have planned to stay in their homes. In fact, S's proposal implies that the only thing the homeowners have to give up is the option to invest the required €5,000 in a different opportunity than the solar panels (a new TV set, say, or an extra holiday). However, deciding to buy the solar panels means reducing the balance in their savings accounts by €5,000 now in order to receive the same amount back over eight years through the decrease in their monthly energy bill. Living up to their obligation to contribute to sustainability *collectively* means that, as an *individual,* each homeowner has to execute the option to invest €5,000 in nine solar panels.

At first, the reaction to S's proposal is simply silence. Nobody seems to know what to say, how to react. The facial expressions of the participants to the discussion change from lively and interested to austere and serious. After a long pause, during which S becomes increasingly overwhelmed by feelings of alienation and uncomfortableness, in a moment of sublimated confusion one of the other homeowners utters the words "What's in it for me?"

At this point, there is some trivial interruption (coffee being served), which provides a convenient opportunity for everybody to drop the topic. After a while, all the homeowners have forgotten at least this part of the meeting, in part because future generations are mute and have no way of getting in touch to ask what is going to be done to respect their interests.

For S, however, this story is not yet over. One way or another, the way the other homeowners have treated her, and her proposal, has had an effect on her that she cannot easily overcome. After all, she wholeheartedly proposed accepting the offer for installation of the solar panels. She was confident that making the proposal to buy the solar panels, after having accepted that we ought to contribute to sustainable development, would be seen as entirely reasonable. S had to conclude, however, that the other homeowners did not share her conception of what was reasonable to do in this situation. The rejection of her initiative has had an emotional impact on her.

It is this experience of S, and the entire situation in which it was evoked, that will be reflected upon in this book. Obviously, S's situation can be understood in many different ways. For example, I do not interpret her situation as simply a cognitive disagreement about, say, whether Minsk is the capital city of Belarus, with the other homeowners insisting that it is Ukraine. A more plausible interpretation could be that people have different conceptions of what it means to live up to a moral duty. They may draw different conclusions about the way moral obligations influence their course of action as individual agents. As a further interpretation, however, it seems to me that S experiences a kind of difference with other people that she has not experienced so far in other situations with the same people. This difference, as expressed by the words "What's in it for me?", brings to the fore the different motives of agents when they agree about the same moral requirement in a given situation. In this book, I will study the motives of individual agents for acting in ways that aim at promoting sustainability.

What does it do to S when other people do not conclude in the way she does that action is required in order to live up to the collectively accepted moral claim that we ought to reduce our carbon footprint? Conversely, what happens to us when we, despite agreeing that it is morally required of us to contribute to a reduction of climate change as a collective, remain sceptical about our contribution as an individual, which we express by saying "What's in it for me?" What happens at the moment that S feels uncomfortable about the acceptance of her proposal to the way she understands herself as an acting human being? What is the effect on us if we, not just once but structurally, persist in a sceptical attitude towards our individual contribution to sustainable development expressible in the words "What's in it for me?"

## 1.2 Understanding sustainability and sustainable action

In discussing the case of the homeowners above, I have tried to highlight some of the many, perhaps confusing, aspects of what it means to take action so as to promote a sustainable future for the planet. The question on which we focus is: Why is it so hard to take such an action? Given that we have problems in finding ways to, for example, mitigate climate change, we need to know more about the obstacles that keep us from engaging in this kind of action. Before we can investigate these obstacles, however, we need to be more specific about what is meant by "a sustainable future for the planet". Once we have formulated a view on the sustainable future we want to strive for, we can be more specific about the sort of action that is needed to get us there.

We will, thus, start with choosing a position on how to understand the concept of sustainability. The concept of *strong sustainability*, as it has been developed recently as a successor to 'sustainable development', gives us a perspective on the future of the planet that seems sufficiently robust to provide normative guidance for human agents. It can also serve as a basis on which institutions can develop their policies. We will then formulate a concept of action that promotes strong sustainability. This is what we will call *sustainable action.*

Although the concept of sustainable development has a history that goes back to the eighteenth century, it entered widespread usage when the Brundtland Commission, also known as the World Commission on Environment and Development (WCED), issued its 1987 report, *Our Common Future.*[1] The report is most famous for the following definition:

> Sustainable development is development that meets the needs of the present without compromising the ability of future generations to meet their own needs. It contains within it two key concepts: The concept of 'needs', in particular the essential needs of the world's poor, to which overriding priority should be given; and the idea of limitations imposed by the state of technology and social organization on the environment's ability to meet present and future needs.
>
> (WCED 1987, 43)

This definition was considered radical, on the one hand, as it established an explicit link between environmental and developmental issues. The report argued that ecological sustainability could not be achieved unless the problem of global poverty was successfully addressed. Both had to be resolved simultaneously and in a mutually reinforcing way. On the other hand, the definition included a reformist element in emphasising a strongly human-centred approach to development within ecological boundaries. This allowed, for example, for a political interpretation in terms of more, not less, human development (Robinson 2004, 372).

Nevertheless, sustainable development has been the subject of criticism. In his assessment of the critique, John Robinson (2004, 373–8) concludes that sustainable development has been interpreted mainly in an instrumental way, meaning that it has been used for other purposes not necessarily related to achieving sustainability. On the level of human values, sustainable development has not sufficiently steered towards greater alignment and unity of interpretation. The concept has been associated too much with technical solutions ("fixes"), instead of providing a normative picture of where the world should go, leading to socially acceptable policies to implement sustainable development.

Therefore, I follow Robinson's proposal to choose the key concept of *sustainability* as the historical successor of *sustainable development*. I understand sustainability as a normative, integrative concept: it guides human action. Sustainability tells us *what* we have to do to live up to our duties, it does not tell us *why* we have to do this. I postpone a discussion of the reasons why we have duties to promote sustainability to §1.3. In line with the WCED report, sustainability summarises our moral duties to future generations. These duties are integrated with our duty to resolve global poverty. Moreover, sustainability imposes limits on human economic activity in order to preserve an amount of natural capital that must be kept constant (SRU 2002, 21).

In the literature a weak and a strong version of sustainability are being distinguished (SRU 2002; Ott and Döring 2008). The main feature of the concept of weak sustainability is that natural capital (the atmosphere, the ozone layer, the carbon cycle, fisheries, forests, rivers, flora, ecosystems, the sea, land, mineral resources, etc.) can be substituted by non-natural capital such as money or scientific knowledge. For example, it would be allowable under weak sustainability to destroy a forest in the expectation that this decrease in natural capital would lead to an increase in monetary or human capital from which future generations would benefit.

Strong sustainability, on the other hand, does not allow replacement of natural capital by manufactured capital; instead these forms of capital are considered complementary (Ott and Döring 2008, 146). This means that some natural capital stocks are incommensurable and non-substitutable, and must be maintained independently of the growth of other forms of capital. If one takes freedom of choice as a normative criterion, then strong sustainability creates more difficulties for current people to implement policies that promote sustainability, as it restricts their options for action. However, it leaves future people more options to make the choices they prefer than weak sustainability does (Ott and Döring 2008, 167). Human economic activity uses resources from nature, which it cannot always reproduce. Strong sustainability implies that human economic activity ought to remain *within* the boundaries set by nature and its capacity to reproduce what economic activity has withdrawn. The amount of natural capital that can be used for economic activity is understood as a theoretical quantity that cannot yet be determined empirically. The natural capital that can be extracted by economic activity depends on the objective of such activity, which could be justified on the basis of a theory of environmental ethics (see §1.3). The crux of the concept of strong sustainability, however, is the *finite* amount of natural capital that exists on earth, including the non-substitutability of its components by non-natural capital (SRU 2002, 64).

On the basis of the problems weak sustainability faces (SRU 2002, 59–64; Ott and Döring 2008, 127–36; Caney 2014) and the potential of strong sustainability to provide normative guidance to agents and institutions, I choose the strong version of sustainability over the weak version. I see strong sustainability as a normative concept that prescribes the duties of current people to future generations, which are integrated with duties to current generations in the sense of fighting global poverty, and the duty to preserve natural capital. There is obviously a lot to say about what sustainability means at the level of concrete human actions. As a minimum, I take it that "the combination of irreversibility and uncertainty should make us more cautious about depleting natural capital. … Some components of natural capital are unique, and their loss has uncertain and potentially irreversible effects on human wellbeing" (Atkinson *et al.* 1997, 16).

In order to provide a more concrete understanding that can be referred to throughout this book, we turn again to Konrad Ott and Ralph Döring, for whom sustainability means that current people must protect various forms of natural capital.[2] Moreover, current people are under the obligation to compensate for their usage of natural capital by an equivalent investment in the creation of natural capital (Ott and Döring 2008, 169). This leads me to adopt the following rules for sustainability at the level of individual agents and institutions:

> Renewable resources ought to be used to the extent that they can regenerate themselves; non-renewable resources and fuels ought to be used to the extent that in parallel to their usage physically and functionally equivalent replacement by renewable resources is realised; polluting emissions must remain within the carrying capacity of the environment and ecosystems, and emissions of non-biodegradable pollutants ought to be minimised, irrespective of whether there is free carrying capacity available.
>
> (SRU 2002, 67)

Through the above rules, sustainability imposes limits on human economic activity in order to preserve an amount of natural capital that forms a fair bequest package for future generations.

Before we can define what a sustainable action is, we need to say a few words about the general concept of action. For the sort of ethical purposes involved here, inadvertently moving one's arm, for example, does not count as an action in the relevant sense. Action is not merely a piece of behaviour that happens to agents. Instead, the concept of action presupposes that the agent aims at a certain goal, and that she at least has the possibility of doing otherwise – that is, of refraining from doing the action, or of doing some other action (Düwell 2014, 35).

The agent voluntarily chooses to act as she does, without being forced by anything other than her own reasons for acting. But voluntariness is not enough: one can voluntarily do things without having a purpose. The sort of action that interests us is also intentional or purposive. The agent acts for some end or purpose that constitutes her reason for acting; this purpose may consist in the action itself or in something to be achieved by the action, which the agent has evaluated as good, and which therefore guides her behaviour (Kalis 2009, 51). We assume that the agent is capable of reflecting on the purposes for which she acts, even if she does not always do so. In performing an action, the agent is aware during the stages of acting that she does the action for the end she has chosen.

This conception of intentional action has formed the starting point for philosophical theories of action ever since it was put forward in different ways by Anscombe (1957) and Davidson (1963). However, in order to be more precise about whether actions are sustainable, I make use of terminology introduced by Christine Korsgaard, who proposes that the constituent parts of the description of an action are the act and the end for which the act is done. This distinction within the concept of action allows consideration of actions in the form of a maxim: I will do act X in order to promote end E (Korsgaard 2009, 11). In other words, the agent does X while she intends to promote end E. This allows us to define the concept of a sustainable action as follows:

***Sustainable Action***

*If we define an action to consist of an act and an end the agent intends to promote, then a sustainable action is an action that fulfils the following requirements:*

1 *The agent chooses the most sustainable act from a range of possible acts at the agent's disposal; and*
2 *The agent's intention is to promote the end of sustainability.*

*Thus, an agent's action is labelled 'sustainable', if she intends to achieve a sustainable end, and her act is the most sustainable act in terms of its consequences given the agent's other options for acts at the moment of acting.*

The end of a sustainable action always forms part of the overall end of sustainability, as outlined above. The end of a sustainable action shall be understood as an individual agent's (or an institution's) intended goal of contributing to sustainability. However, individual agents or institutions alone cannot reach the goal of sustainability. The concerted effort of the human community as a whole, including its institutions, is a necessary condition for sustainability to be realised. For example, contributing to the end that fresh water will be preserved (e.g. by reducing one's meat consumption) – because climate science predicts that fresh water will become a scarce resource in a few decades from now – can be considered a sustainable end. To make this happen, many other agents will need to take similar actions, and thus institutional policy implementations (e.g. communication campaigns) are necessary.

The definition not only requires the end of a sustainable action to be sustainable, it also requires the act itself to be sustainable. This means that from a range of alternative acts that are at the disposal of the agent, by which the same end could be achieved, the act that is performed must be optimal in terms of its contribution to sustainability. To continue with the example of preserving fresh water: since science shows that the amount of fresh water needed to produce one kilogram of meat is much greater than what is needed to produce a crop product with equivalent nutritional value (Mekonnen and Hoekstra 2010), reducing one's meat consumption is the preferred act to contribute to the end of preserving fresh water for future people. Similarly, commuting to the office by public transport instead of by car will be the right act for contributing to the end of reducing one's greenhouse gas emissions, thereby reducing the population's greenhouse gas emissions. In summary, the act involved in a sustainable action is the most favourable one from a set of alternative acts within reach of the agent at the moment of acting in terms of its consequences for sustainability.

As stated earlier in this section, we understand sustainability as a normative concept that imposes limits on human economic activity in order to preserve an amount of natural capital. Since agents are required to follow the rules of sustainability, they have at least a prima facie duty to choose the most sustainable act from a range of possible acts at a moment of acting. However, this does not mean that any specific evaluative claims are made when agents and their actions fail to meet the standard of "sustainable action". There can be good reasons why someone takes the car to the office instead of using public transport, and still wants to realise a sustainable future. There can be equally good reasons not to reduce one's meat consumption, while still contributing to sustainability in another way. The purpose of the concept of sustainable action is to look at the agent's aims, intentions, and purposes – that is, the end she wants to achieve with her act – *and* to ask whether at the moment of acting there are other more sustainable acts at the agent's disposal for which she could have opted, as her action is voluntary.

We can continue with identifying more nuances in order to distinguish between actions that promote sustainability and actions that do not. The latter actions are called *non-sustainable* if the agent does not intend to contribute to sustainability and the act itself is not optimal in the sense that there are more environmentally friendly acts available for the agent at the moment of acting. We can also distinguish actions that are done to achieve a different end than sustainability, but that are nevertheless carried out in the most sustainable manner according to what climate science prescribes at the moment of acting. I call these *instrumentally sustainable* actions. Moreover, it is possible to have actions that are done for a sustainable end, but that are not favourable as regards their contribution to sustainability. I label such actions *suboptimal.* A matrix for the different forms of sustainable action is represented below.

To illustrate, let us consider examples of actions that are not sustainable in the sense that the agent intends to achieve a sustainable end, *and* her act is the most sustainable act in terms of its consequences given the agent's other options for acts at the moment of acting. For example, acting sustainably with the intention to increase a company's profit can be called an instrumentally sustainable action. An example of a suboptimal sustainable action would be the manager of a company fully dedicated to producing environmentally friendly products commuting by car to the office over a small distance, whereas there is convenient public transport available. In a similar vein, former US vice president Al Gore Jr taking aeroplanes frequently to raise awareness of climate change by delivering his lecture about *An Inconvenient Truth* could also be considered a suboptimal sustainable action. Again, the intention is not to morally judge Al Gore for not doing otherwise. From his point of view, all things considered, it was probably most effective to achieve political momentum in favour of sustainability to take the plane. Instead, I want to emphasise the phenomenon that many times we intend to act sustainably, but circumstances force us to choose a suboptimal act. The intention to contribute to sustainability is there, but we simply see no possibility to act on it, or other reasons prevail that make us act suboptimally.

Whereas the case of suboptimal sustainable action seems reconcilable with how we experience the contingency of our lives, one might ask whether there is a meaningful difference between sustainable and instrumentally sustainable actions, since both involve a sustainable act and only differ with respect to the

Forms of sustainable action

| | | *Does the agent intend to contribute to sustainability?* | |
|---|---|---|---|
| | | *Yes* | *No* |
| *Is the agent's act sustainable?* | Yes | Sustainable action | Instrumentally sustainable action |
| | No | Suboptimal sustainable action | Non-sustainable action |

agent's intention. Obviously, there are many cases of instrumentally sustainable actions, for example when we act sustainably out of habit, because of peer pressure, or because we are carrying out an assignment. There is, however, empirical evidence suggesting that indeed there is a meaningful difference between instrumentally sustainable and sustainable actions.

First, empirical studies reveal that the energy saving behaviour of people who act on the moral motivation to do something for the environment is more robust in resisting temptations to revert to prior behaviour than the behaviour of people who act for hedonic or economic reasons. People who are motivated by environmental concern are more likely to continue with their energy saving behaviour than people who do this out of economic interest, even when this becomes less attractive or the cost argument no longer applies (Steg 2008). This observation is supported by a study where the motivation of people to comply with environmental policies was investigated. It was found that appeals to economic motives elicit less compliance than an appeal to environmental concern (Bolderdijk *et al.* 2012).

Second, the so-called rebound effect seems different for instrumentally sustainable and sustainable actions. The rebound effect entails, for example, agents insulating their homes in order to save energy, then rewarding themselves by switching on the heater for a longer time. Thereby, the effect of reducing one's energy consumption through home insulation is partly eliminated (Aydin 2016). Likewise, people who install a water-saving shower device reward themselves by taking a longer shower. In short, the rebound effect emphasises that the savings of an energy efficient product are partly eliminated, since the agent uses it less sparingly than she could have done. For the ex ante assessment of the effect of institutional energy-saving policies, the rebound effect thus seems of crucial importance. The net effect of any policy may be lower due to the rebound effect.[3]

Taken together, these empirical findings suggest that the rebound effect, which seems empirically inevitable, could be lowered when policies for energy saving emphasise the normative and environmental concerns instead of the economic argument of saving money. This means that we have found a significant difference between instrumentally sustainable and sustainable actions in the sense of their contribution to achieving sustainability.

Finally, to illustrate non-sustainable action, I borrow an example from Walter Sinnott-Armstrong. Since this example will be discussed later, in §2.2, I cite it here in its entirety.

> My example will be wasteful driving. Some people drive to their jobs or to the store because they have no other reasonable way to work and eat. I want to avoid issues about whether these goals justify driving, so I will focus on a case where nothing so important is gained. I will consider driving for fun on a beautiful Sunday afternoon. My drive is not necessary to cure depression or calm aggressive impulses. All that is gained is pleasure. Ah, the feel of the wind in your hair! The views! How spectacular!

> Of course, you could drive a fuel-efficient hybrid car. But fuel-efficient cars have less "get up and go". So let us consider a gas-guzzling sport-utility vehicle. Ah, the feeling of power! The excitement! Maybe you do not like to go for drives in sport-utility vehicles on sunny Sunday afternoons, but many people do.
>
> (Sinnott-Armstrong 2010, 333)

As I understand sustainable action, we have to look at the combination of the act *and* the end to see if this gives us grounds to conclude that the example of the Sunday afternoon drive in a gas-guzzler is non-sustainable. It is quite obviously so: the end is pleasure, not sustainability, and the act is deliberately not the most sustainable one given the end of a pleasurable experience. Still, this example gives rise to two useful considerations.

First, the act of the Sunday afternoon drive positively contributes to climate change. Empirically speaking, the effect of this act is that a certain amount of greenhouse gases is emitted. This emitted quantity of greenhouse gases would not have been there if the drive had not been taken. Since climate change is caused mainly by greenhouse gas emissions, the drive in the gas-guzzler makes a contribution to climate change, however small it may be. But this act also contributes theoretically to a deterioration of the climate. Human beings are discrete – that is, they are individually separate and distinct – and their number is finite. Each human being emits a finite number of $CO_2$ molecules into the air, which means that although the contribution of the drive to the climate change problem is small, it is by no means zero, and the number of $CO_2$ molecules in the atmosphere would have been lower if the drive had not been taken.

Second, although we may claim to have a liberty right to decide to go for a drive in a gas-guzzler on a sunny Sunday afternoon, exercising this right seems problematic. As the example is construed, the drive is not necessary to reach the workplace or a supermarket, nor is it necessary for any reason concerned with the driver's psychological well-being. Hence, although we may argue that we have the right to cause emissions that keep our quality of life above a minimum level, the drive can safely be considered one that is not necessary for our subsistence; the drive causes so-called luxury emissions (Shue 2010; Peeters *et al.* 2015, 28). Since we must adhere to the norm that, collectively, the affluent countries should reduce their environmental footprint substantially and rapidly enough to allow the developing countries to build up a just level of wealth, whilst simultaneously global environmental damage must decrease according to, for example, the projections of the Intergovernmental Panel on Climate Change (IPCC), the claim to take this drive on the basis of a moral right cannot be justified.

In conclusion, the drive in a gas-guzzler does not meet the criteria of a sustainable action. Since the act of driving leads to a further deterioration of the climate, *and* the end of the action is incompatible with sustainability, the drive in a gas-guzzler on a sunny Sunday afternoon qualifies as an example of a non-sustainable action.

## 1.3 The need for a philosophy of motivation for sustainable action

Having elaborated the example of S, who attempts to accomplish a sustainable action in a social context, and having clarified how I understand sustainability and sustainable action, we now consider what philosophy can contribute to understanding the question: *Why is it so hard to engage in sustainable action?* To focus the discussion, I will first distinguish three domains in the sustainability debate. I will then give a brief overview of those domains that lie *outside* the scope of the book: the epistemic or scientific domain, and the domain of normative ethics. The main contents of these three domains are summarised in a condensed statement of the challenge that humanity currently faces. This challenge provides the background for the topic of the book: the third domain, which is the domain of human motivation. I argue that, as philosophers, we need to understand better the particular problems surrounding an agent's motivation for sustainable action.

The global situation regarding climate change poses a problem for humans alive today that transcends numerous boundaries. Due to its global nature, climate change transcends national boundaries.[4] Due to its intertemporal nature, climate change goes beyond generational boundaries (*The Guardian* 2012). Finally, climate change challenges political and scientific theories that were believed to be robust enough to cope with the environmental problems the world is facing. An example of a theoretical challenge is the economic theory of cost-benefit analysis. Cost-benefit analysis suggests that one can compare scenarios for mitigating or adapting to climate change through using discounting techniques. However, from a moral point of view, discounting cannot be used as a justification for the obligations of current generations towards future generations. Climate change thus also transcends conceptual boundaries.[5]

Within the field of practical reason, in which one deliberates about one's actions from a first-person point of view (Wallace 2014), we need to understand how our actions as individuals relate to the impact of our collective behaviour on the climate. As we saw in the case of the homeowners (see §1.1), even if we accept the normative reasons (reasons that justify our action) in favour of taking sustainable action, our motivating reasons (reasons that explain our action) influence our intentional actions as well. Although there was agreement about the normative reasons, there appeared to be differences regarding the motivating reasons of these homeowner agents. As practical philosophers, we thus need to reflect on how our motivating reasons influence our course of action given the exceptional challenges imposed by climate change. That is, my investigation focuses on our *motivating* reasons for taking sustainable action as highlighted by the question: Why is it so hard to engage in sustainable action?

In order to establish a context within which this question can be discussed, I will argue for two assumptions that are necessary for outlining the topic of the book. The first relates to the debates that are ongoing about the scientific evidence for the existence of anthropogenic climate change. In order for anyone to know that there is a problem with climate change, there must be scientific evidence that climate change is happening, and as to what the consequences of inaction will be. That is, before anything can be said about duties to mitigate climate change, the current way of living must be problematised by providing knowledge about its consequences. This evidence is globally collected and synthetised by the IPCC. This means that we have to take seriously the science of climate change, and the institution that interprets climate science and conveys its results. As I will argue below, after having received the fifth IPCC report in a row, we are left in hardly any doubt that climate change occurs, and that its main cause is anthropogenic.

My second assumption relates to the debate in normative ethics as to whether or not we have an obligation to mitigate climate change. If we consider natural science reliable enough to inform us about the causes and effects of climate change, then the question arises whether we have any obligation to mitigate a global temperature rise. In other words, on the basis of the scientific reports that climate change is happening, an argument has to be found for the norm that the world population has an obligation to take action to mitigate climate change. Accordingly, as the science of climate change has advanced, the debate about the normative aspects of climate change has arisen. In this section, I will discuss two families of normative theories. I will conclude that, for a study of the problem with motivation for sustainable action, it does not matter which group of theories is preferred as the normative justification for mitigating climate change.

Finally, having presupposed the norm that, as a collective, humanity has the obligation to mitigate climate change, which in turn relies on the results of climate science, we see the motivation problem starting to kick in – the focus for this book. I view the motivation problem as a multi-faceted problem: it can be characterised by questions that play a role in different contexts. Empirical questions that arise are: Why do climate science and the norm to mitigate climate change fail to morally motivate us to collectively reduce our greenhouse gas emissions? And, what makes sustainable action distinctive when it comes to motivating the individual agent? However, given that changes in the behaviour of the public do not go fast enough to meet the greenhouse gas emission reduction targets, the question as to what guidance institutions ought to give to individuals in order to act sustainably is an ethical one; and the question as to what we realistically can expect from institutions in this regard is an empirical one. Finally, it is an anthropological-ethical question how our practical self-understanding is affected if humanity structurally and persistently fails to get its act together regarding the mitigation of climate change.

In conclusion, three domains can be distinguished within the sustainability debate: the epistemic or scientific, the normative ethical, and the motivational. These domains need to be addressed in this order. That is, without knowledge delivered by climate science there is no ground for a normative requirement to mitigate climate change,[6] and without a normative requirement we can hardly start thinking about problems with motivation for sustainable action. I will now continue with the discussion of these domains, as this allows for a clear localisation of our topic.

As stated earlier in this section, the first presupposition of this book is that we should consider climate science prima facie reliable in informing us about the effects on the climate of the collective behaviour of human beings. Although debate about the IPCC and its Assessment Reports is ongoing, I give three reasons for arguing that the messages of this body warrant to be considered as a basis for the normative requirement that, as a collective, humanity must act to mitigate its impact on the global climate.

First, from the start the IPCC has been central to the assessment, validation, and mobilisation of knowledge claims about climate change (Hulme and Mahony 2010, 706). The IPCC was founded by the United Nations (UN), giving it authority and the assurance that the UN members would consider its findings (UN 1988). By means of its organisation into separate working groups allocated to different tasks (science of climate change, study of social and economic impact of climate change, and response strategies to limit or mitigate the impact of climate change), the IPCC endeavours to live up to its UN mandate. In the UN climate negotiations that have taken place since 1992 (Rio de Janeiro), the findings of the IPCC, as communicated through its Assessment Reports, have been embedded in the UN Framework Convention on Climate Change (UNFCCC).

Not surprisingly, this political influence has resulted in criticism of the scientific independence of the IPCC. Jeroen van der Sluijs *et al.* argue that the political role of the IPCC to date has not received enough attention. The authors discuss the way in which this role has had an effect on issues such as quantifying uncertainty, building scientific consensus, and dealing with openness and ignorance. After discussing the weaknesses of the consensus approach, as followed by the IPCC, the authors conclude that climate science should become less political, "by offering room for dissent within climate science and communicating about it with policymakers" (Van der Sluijs *et al.* 2010).

Second, although the debate is continuing, it seems that consensus among climate scientists about the anthropogenic causes of climate change has become stronger over the years (see Bray (2010) for a sociological approach). The findings of some of the main studies about the consensus among climate scientists are as follows. Naomi Oreskes reviews 928 papers, of which 75 per cent support the consensus view, and 25 per cent take no position on anthropogenic climate change; however, none of these latter papers reject the consensus view. She addresses the communication problem of scientists by concluding that despite the consensus among them, journalists and politicians tell the public that there is still confusion and disagreement among scientists (Oreskes 2004). Peter Doran and Maggie Kendall Zimmerman perform a survey among over 3,000 climate scientists, mainly from the US Of these respondents, 90 per cent think that mean global temperatures have risen since pre-industrial levels, and 82 per cent think that human activity has been a significant contributing factor (Doran and Kendall Zimmerman 2009). Subsequently, John Cook *et al.* examine 11,944 climate science abstracts from the period 1991–2011. They conclude that the number of papers rejecting the consensus on anthropogenic global warming is a "vanishingly small proportion of the published research" (Cook *et al.* 2013).[7]

Third, since its commencement, the IPCC has issued five Assessment Reports. The main claims of Working Group I formulate the overall changes in the climate and provide causes thereof. Over the years, these claims have increased in clarity, accurateness, and certainty (IPCC 1990, xi; IPCC 1992, 5; IPCC 1995, 3; IPCC 2001, 2, 10; IPCC 2007, 2, 5; IPCC 2014a, 3, 4). As Eleftheria Vasileiadou *et al.* argue, the IPCC has changed its qualitative approach in the second Assessment Report (1995) ("the balance of evidence suggests a discernible human influence on global climate") to a probabilistic expression in the third Report (2001). The "likely" (>66 per cent chance) in the third Report increases to "very likely" in the fourth Report (2007) (Vasileiadou *et al.* 2011, 1059), and further to "extremely likely" (>95 per cent chance) in the fifth Assessment Report.

To summarise, the IPCC has been criticised for applying a consensus approach and for not allowing for more openness about dissent within the scientific community, and also for its methodology of assessing consensus among climate scientists. Despite the critique, as the above considerations make clear, there seem sufficient grounds for the plausibility of the underlying assumption of this book that there is a likelihood of over 95 percent that anthropogenic drivers have been the dominant cause of the observed warming since the mid-twentieth century.

The second presupposition of the book is the norm that present human beings collectively have an obligation to prevent a global temperature rise of more than 2 °C. Below, I provide an overview of two ethical theories on the basis of which this normative requirement can be justified. These theories are members of two respective families of theories that have emerged in environmental ethics: one that takes the view that our moral duties towards the environment are derived from our direct duties to its human inhabitants, and one that puts into the centre the intrinsic value of nature (Brennan and Lo 2015). From the former group of theories (Passmore 1974; Bookchin 1990; Norton *et al.* 1995), I will give a short account of the theory of *intergenerational justice*. The theories in this group take the duties of current people towards future people as a starting point.

However, from the concept of sustainability (see §1.2) a further question emerges: What exactly does sustainability seek to sustain? Theories from the latter group (O'Neill 1992; Jamieson 2002) have attempted to answer this question. This family of theories argues that referring solely to the interests of current and future people is inappropriate, since nature has its own moral status that should be respected. I will, therefore, as an illustration of this group of theories, give some insight into the theory that argues for the *intrinsic value of nature*.[8]

The following are some of the crucial elements of the concept of intergenerational justice.[9] First, there is debate about the extent to which complicated forms of collective action and social ontology have to be presupposed in order for (future) generations to be the bearer of rights – that is, to be the bearer of rights as a generation. For this reason, when discussing intergenerational justice, reference is made to (future) people instead of generations as cohorts of human beings. Moreover, as we are dealing with the (possible) rights of future people and the duties that follow from those rights for current people, current generations, too, should be understood as a multitude of current individuals, not as a group. In this book, I will thus continue to use the term 'intergenerational', but with the meaning of current or future people, or current or future persons (Unnerstall 1999, 126).

Second, we should be more precise about the time value of the rights of future people. In order for a bearer of rights to be able to exercise them, she has to be actually present in space and time. This implies that we cannot speak of future people currently having rights, since future people do not yet exist. In the literature on intergenerational justice, the consensus is that we can speak in a meaningful way only of the future rights of future people (Unnerstall 1999, 52). Thus, so far we can say that theories of intergenerational justice attempt to answer the question: Which duties do current people have in order not to violate the rights future people will have in the future?

When we think about the rights future people will have in the future, we are dealing with a set of claims that someone can uphold against others on the basis of certain conventions, laws, norms, or principles. In an ethical context such as debates about intergenerational justice, we are especially interested in moral rights. Moral rights are understood as claims that human beings may legitimately address to each other. They are 'moral' rights insofar as their justification does not depend on the fact that they are actually implemented in the law. Nevertheless, moral rights are not independent of a legal dimension to the extent that they provide strong reasons for the necessity of creating legal regulations that are appropriate ways of protecting them. Moral rights, and more specifically moral claim-rights, appear in the following form: A has with respect to B a right to X on the basis of Y. In this definition A is the subject or bearer of the right, and B the addressee of the right – that is, the one for whom the obligation counts to respect the right to which A is entitled. X expresses the right's content, and Y constitutes its legitimate foundation (Hübenthal 2007, 283).

If we try to clarify the components of this definition, then we have so far spoken about A (i.e. future people) and B (i.e. current people). It is already an extension of the concept of moral rights that A and B are not simultaneously present, but that A appears long after B has disappeared. Furthermore, the content X specifies what it is to which future generations are entitled. At this point we should realise that the debate about sustainability is mainly concerned with determining this content. For example, as discussed in §1.2, we distinguish here between weak and strong sustainability, preserving natural capital, and so on (SRU 2002; Ott 2004). As it is beyond the scope of this book to go into detail regarding the possibilities for X, consistent with the Universal Declaration of Human Rights (UDHR 1948), I have chosen for the time being the following general formulation: future people have, with respect to current people, the right to a life as a human being.[10]

So far, we have found the following formulation of intergenerational justice: future people will have with respect to current people the right to live their lives as human beings on the basis of an ethical theory Y that needs further specification. Consequently, current people have the duty to respect the right of future people to live their lives as human beings.

What are the consequences of this duty for current people? One consequence is that current people should not take actions that can result in a violation of the right of future people to their life as human beings. That does not imply that we have a duty to bring future people into existence, but it implies the duty not to infringe their rights if they should come into existence, and this forms a duty for us to the extent that we have reasons to assume that there will be future people. And since actions by current people affect the well-being of future people, and thus have an influence on the degree to which future people are able to exercise this future right, the duty mentioned here counts as a moral duty. The scenario of future people having a right to a life as a human being is thus sufficient for the foundation of normative consequences that have a bearing on the actions of current people (Unnerstall 1999, 450).

In summary, a theory of intergenerational justice argues for the moral claim that current people are under the obligation to take appropriate measures, or to accept limits on their options for action legitimately imposed by an institution (e.g. a sovereign state), in order to respect the (future) right of future people to live as a human being. We will now consider a member of the family of theories that justifies a normative requirement to mitigate climate change on the basis of the intrinsic value of nature.

If natural objects and systems have intrinsic value independent of their possible use for human beings, then a policy approach to sustainability needs to consider the environment and natural things not only in instrumental but also in intrinsic terms to do justice to the moral standing that people believe such items to possess. Because the intrinsically valuable is that which is good as an end in itself, it is commonly agreed that something's possession of intrinsic value generates a prima facie direct moral duty on the part of moral agents to protect it or at least refrain from damaging it (Brennan and Lo 2015).

Theories about the intrinsic value of nature emphasise that human beings are also a part of nature. These theories particularly aim to demonstrate the deficiencies of a view that is only focused on a short-term instrumental use of nature that does not entail a sensate, aesthetic, and emotional relationship to nature. For example, deep ecology endorses "biospheric egalitarianism". This is the view that all living things are alike in having value in their own right, independent of their usefulness to others (Næss 1973). It is presupposed that humans have a duty directly to species, in addition to obligations to other humans. A comprehensive environmental ethics reallocates value across the whole continuum of humans, animals, plants, species, and ecosystems. When humans value, they must not forget their communal bonds to these parts of nature (Rolston 1995, 1998).

A debate that falls within this group of theories is about strong sustainability, which does not allow for substitution of natural capital by human or manufactured capital (see §1.2); instead these forms of capital are considered complementary. This means that the interpretation of 'capital' in natural capital is different from how it is interpreted in economic science, namely as a resource to be utilised to produce a benefit for human beings, which they can command. It is difficult, however, to define natural capital, since its components are closely interconnected. Simply producing a list of differentiated, seemingly well-demarcated elements of natural capital seems to miss the interconnectedness of these heterogeneous parts, each of which performs a different function in respect of humankind. That is, natural capital is characterised by terms such as 'biodiversity' and 'stability of ecosystems'. Although there is wide consensus that the elements mentioned in §1.2 form part of it (the atmosphere, the ozone layer, the carbon cycle, fisheries, forests, rivers, flora, ecosystems, the sea, land, mineral resources, etc.), the concept of natural capital is nowhere developed beyond this schematic level (SRU 2002, 64).

From their experience of nature, human beings obtain a broad spectrum of eudaemonistic values (Ott and Döring 2008, 205). In a global assessment of ecosystems, nature is described as providing "ecosystem services" to human beings in the form of provisioning (food, fibre, genetic resources, biochemicals, and fresh water), regulating (air quality, climate, water, erosion, etc.), and cultural services (spiritual and religious values, aesthetic values, recreation and ecotourism) (Millennium Ecosystem Assessment 2005, 7). This means that our duty to protect natural capital can be based on the eudaemonistic reason that human beings need nature for their own well-being.[11]

The question that is of interest for this book is whether theories of intergenerational justice and the intrinsic value of nature differ from each other when it comes to how they motivate the agent towards acting sustainably. In making the choice to act sustainably, would a theory of intergenerational justice affect the reasons that explain my action – that is, my motivating reasons – in a different way from a theory about the intrinsic value of nature? For example, when choosing between different means of transportation (bicycle, car, aeroplane, etc.), would it matter whether I do this with the intention to protect nature, or with reference to my duties to future generations?

As I will argue below, when discussing the domain of human motivation, whatever normative theory we develop to justify sustainable action (i.e. the theories that have been discussed above in this section), such theory will have problems regarding its motivating potential. In other words, the problem with motivation for sustainable action does not disappear as a result of finding a normative theory with which such action could be justified. In the course of this book, therefore, I consider both groups of theories to be equally capable of providing a normative justification, and I remain agnostic as to which group of theories would be preferable for dealing with the problem with motivation for sustainable action. However, in order to be clear about the content of the moral requirement to mitigate climate change, I choose in favour of intergenerational justice as supporting normative theory.

On the basis of the two presuppositions discussed above in this section, I continue with formulating the collective action problem that humanity currently faces. This action problem in the form of a condensed statement serves as the starting point for a discussion about the need for a philosophy of motivation for sustainable action. Inspired by an example from Stephen Gardiner, I formulate this challenge as follows:

***Humanity's Challenge***

*"Humanity is, in geological and evolutionary terms, a recent arrival on the planet, and is currently undergoing an amazingly rapid expansion, in terms of sheer population numbers, technological capabilities, and environmental impact" (Gardiner 2011b, 56). Against this background, the IPCC reports: "Anthropogenic greenhouse gas emissions have increased since the pre-industrial era, driven largely by economic and population growth, and are now higher than ever. This has led to atmospheric concentrations of carbon dioxide, methane and nitrous oxide that are unprecedented in at least the last 800,000 years. Their effects, together with those of other anthropogenic drivers, have been detected throughout the climate system and are* extremely likely *to have been the dominant cause of the observed warming since the mid-20th century" (IPCC 2014a, 4, emphasis IPCC).*

*In response, 195 members of the United Nations ("The Conference of the Parties") are "emphasizing with great concern the urgent need to address the significant gap between the aggregate effect of Parties' mitigation pledges in terms of global annual emissions of greenhouse gases by 2020 and aggregate emission pathways consistent with holding the increase in the global average temperature to well below 2°C above pre-industrial levels and pursuing efforts to limit the temperature increase to 1.5°C above pre-industrial levels" (UNFCCC 2015, 2).*

*A basic question that faces us as humans, then, is whether we can meet the challenge of changing our way of life such that we live up to our commitment to do justice to the people who will live long after us by taking sufficient sustainable action to meet the target of "holding the increase in the global average temperature to well below 2°C above pre-industrial levels ..., recognizing that this would significantly reduce the risks and impacts of climate change" (UNFCCC 2015, art. 2.1(a)).*

We will refer to Humanity's Challenge throughout this book as the main background assumption for considering the problem with motivation for sustainable action. Humanity's Challenge presupposes that the claims of climate science are trustworthy, *and* that the norm to respect the (future) rights of future generations is valid. Hence, the question of motivation for sustainable action becomes relevant after we have accepted Humanity's Challenge.[12] When discussing, from here onwards, the third domain, which is the domain of human motivation, I argue that we need a philosophical understanding of the particular motivational problems with agents taking sustainable action.

In general, the debate about whether a norm is capable of motivating an agent to act in conformity with it is as old as practical philosophy itself. For example, Konrad Ott calls the problem of motivation one of the "structural problems" of applied ethics. Insight into what is ethically required does not *ipso facto* result in sufficient motivation for a suitable course of action. Ethics alone cannot prevent free rider behaviour. Free riders are not moved by the reasons that justify norms, and, consequently, the problem of motivation can be solved only as part of a moral culture that encompasses, amongst other things, ethical theory, moral education by parents and teachers, a free press, laws, and an independent judicial system (Ott 1996, 62–3; Birnbacher 2009, 273). Hence, instead of discussing meta-ethical theories that argue in general for an internalist or externalist view on motivating agents to act in accordance with moral claims, I will argue that a philosophy of motivation is necessary, specifically in the case of human action that is intended to take up Humanity's Challenge.

Other authors have argued that actions that promote sustainability seem vulnerable to overriding motives that pull us in a different direction from the one we cognitively agreed to (Care 1982; Baumgartner 2005; Birnbacher 2009; McKinnon 2011; Kretz 2012; Bourban 2014; Peeters *et al.* 2015). Dieter Birnbacher has done groundbreaking work in identifying and exploring the problem with motivation for sustainable action. He notes that instead of understanding the problem of motivation in terms of bridging the gap between the *acceptance* of a norm and the *action* that follows up on it, a more fine-grained model may be required, with two more action phases in between: *adoption, application*. Thus, Birnbacher conceives of motivation for sustainable action as acting in accordance with a norm, whereby the agent goes through the four phases of acceptance, adoption, application, and action (Birnbacher 2009, 274).

On the basis of this model, Birnbacher argues that our motivation to care for people in the distant future is a special case. In his view, agents have fewer issues with the motivation to accept ethical principles related to preservation of the planet for future generations than their adoption of those principles. Given "a moral culture deeply impregnated with the universalistic moral tradition of the Enlightenment", it seems likely that we judge it right that principles related to our dealings with present people extend to our dealings with future people (acceptance). It is, however, not evident that we let this rule guide our behaviour and that we incorporate it into our own identity (adoption) (Birnbacher 2009, 276).

Birnbacher argues that two factors make the problem of motivation for sustainable action unique compared with other situations where an agent's motivation is challenged. First, actively taking responsibility for sustainability depends more on moral motives, and, second, the effect of an action that promotes sustainability is less certain (Birnbacher 2009, 281). Whatever the reasons that normatively support sustainable action, at the motivational level we have to deal with the potential psychological inconsistency between what practical reason commands and the actions that we in fact carry out. This happens, for example, when we engage in wasteful driving on a sunny Sunday afternoon or, year on year, increase the rate of depletion of natural resources (International Energy Agency 2014), despite the fact that we are perfectly aware that these resources are finite, and that people in the distant future will also need to benefit from them.

I argue that there are four philosophical[13] questions that differentiate the problem with motivation for sustainable action from motivational issues in other morally relevant contexts. It is the interplay between and the accumulation of these questions that is unique for the problem of promoting sustainability. Like Stephen Gardiner, who argues for the "perfect moral storm" as a metaphor for what seems the root cause of the global environmental tragedy, my claim is that the mutual reinforcement of the issues these questions raise justifies the philosophical investigation into the problem with motivation for sustainable action undertaken in this book.

The four questions I have in mind are:

1. *Which meta-ethical presuppositions are necessary for an assumption that human beings should be motivated to take sustainable action?*
2. *What or who should be considered the addressee of the moral duty to take sustainable action?*
3. *What can be expected of the addressee as regards acting in conformity with the norm to meet Humanity's Challenge?*
4. *How can we conceptualise the motivating force of the addressee's obligations to respect the rights of future generations?*

These questions constitute forms of philosophical obstacles that hamper the agent's motivation for sustainable action. Such obstacles need to be taken away before we can start investigating the specific features of motivation for sustainable action and comparing them with the motivational structure of other morally complicated actions. In the remaining paragraphs of this section, I will provide an overview of the arguments that I will elaborate later in the book. Therefore, I keep this overview brief and concise.

According to the first question, the entire enterprise of investigating the motivational problems with sustainable action could potentially collapse at the meta-ethical level. Dale Jamieson argues that the problem of climate change can be considered a non-paradigmatic moral problem, meaning that it does not have the features of cases in which an individual intentionally harms another individual, as in cases such as "Jack intentionally stealing Jill's bicycle" (Jamieson 2010, 436). As a consequence, according to Jamieson, viewing climate change as a problem of individual moral responsibility requires a revision of the concept of moral responsibility (Jamieson 2010, 438). Note that Jamieson identifies normative reasons with motivating reasons (Gardiner 2011b, 41). His view thus implies that, as agents cannot assume individual moral responsibility, their motivation to do something about climate change is hampered. If this indeed be the case, the problem with motivation for sustainable action would need to be approached under a view that considers the agent's motivation as independent from the reasons that justify sustainable action (Gardiner 2011b, 42). Having accepted Humanity's Challenge as a justifying reason, under that view we would not be able to investigate what it means to be motivated to take sustainable action. In reply to this potential objection, I will defend weak internalism as one of the conceptual premises on which the argument in this book relies (§1.4). This is the view that agents feel *some* motivation to act in accordance with their moral judgements; that motivation, however, can be overridden by conflicting desires and defeated by a variety of mental problems, such as depression and weakness of will (Rosati 2016).

The second question relates to the problem of assigning responsibility for taking sustainable action, whether to individual agents, collectives, or institutions. Given the global nature of the sustainability problem, there are various views as to what or who should be considered the addressee of the requirement to work on the mitigation of climate change. It seems obvious that the political institutions have their own responsibility to implement policies that aim to meet Humanity's Challenge. However, I will defend the view that individual human agents in affluent countries have *some* shared responsibility to take sustainable action, irrespective of the institutional policies. This view has the methodological implication of allowing us in the first instance to focus on understanding the problem of motivation for sustainable action from the perspective of individual agents. This is the *first* problem to be investigated. The design and implementation of institutional policies that steer our actions towards more sustainable ones can then make use of the insights thus obtained (§2.2).

The third question comes to the fore once we have concluded that there is a duty for the individual agent to act sustainably. This question refers to how demanding moral duties to act sustainably can be. For example, although there could be a moral requirement that tells me to reduce famine by donating part of my salary, it cannot be required of me to donate a significant part, as this would threaten my own existence, and would be unreasonable, from a common-sense point of view. However, in the case of S, who proposes that she and her fellow homeowners place solar panels on their roofs (see §1.1), it seems to accord with her standard of reasonableness to bear the relatively small burden of investing €5,000 up front and receiving this amount back roughly within eight years by means of a reduced monthly energy bill, whereas this seems unreasonable in the eyes of the other homeowners.

Therefore, given the global nature of the climate change problem, which requires concerted behaviour by many human beings and institutions, one might infer that the level of coordination required is simply too demanding for us. How can we contribute as individuals to meeting Humanity's Challenge given the extremely small impact of our individual actions on the entirety of actions by seven billion people currently populating the planet? I will argue that, in order to become motivated to take up Humanity's Challenge, we need to extend our moral horizon so as to include future people as potential victims of our current non-sustainable behaviour. In particular, institutional help will be needed to tackle this problem (§4.3).

The fourth question deals with the problem that the motivational force of our obligation to act sustainably could be weakened due to the incompatibility between the conception of the human being that we apply to future people and our conception of current people. In other words, our motivation as agents could be hampered by the way we conceptualise future people, since the moral status of these people depends on the conception of the human being we use in our deliberation about sustainable action. To clarify this claim, let us suppose an agent intends to contribute to meeting Humanity's Challenge, and she believes that a particular action is a means to that end. The motivational problem arises if she believes that the identity of future people is fully contingent upon the actions of current people. That is, if future people are born, they will owe their very existence to the actions that current people have taken. Since those actions constitute their identity, they will not regret what current people have done. Hence, it does not matter whether the agent acts sustainably now. I will argue that working towards meeting Humanity's Challenge means that we must commit to act towards that end over decades and centuries. This means that we need an understanding of ourselves and other agents, with whom we cooperate in social institutions, which is based on the continuity of our identity over time. Therefore, I will argue, we must consider ourselves as unified agents over time (§4.4).

As the above considerations make clear, these four philosophical questions have to be dealt with sequentially. That is, the response to the first question determines the relevance of the second question, the response to the second question determines whether the third question becomes meaningful, and so on. If my argument for the relevance and interrelatedness of these questions has been convincing, I take it that we have prima facie reasons to start the philosophical investigation into the motivational features of sustainable action. Following up on the common-sense questions (i.e. why does it seem so hard to engage in sustainable action; and how does such action differ from other morally complicated actions, motivationally speaking?), the more detailed question that will lead us in this investigation is: What differences are there in the relationship between the duty to live up to our obligations to meet Humanity's Challenge and our motivation as individual human agents for sustainable action, and the relationship between another morally relevant norm and our motivation for an action in conformity with that norm?

So far, I have, first, demarcated the field of motivational questions in relation to sustainability from those of scientific and normative questions. Second, I have proposed that climate science is advanced enough to serve as a foundation for the investigation of the motivational problems. Third, I have argued that the normative requirement to mitigate climate change can be justified on the basis of theories of intergenerational justice. This analysis led me to, fourth, formulate Humanity's Challenge as the main assumption that helps in obtaining a clearer view on what the problem with motivation for sustainable action entails. Fifth, I have argued that a philosophical inquiry into this problem is required, since there are four questions to be dealt with that make this problem special compared with the motivational challenges agents encounter in other morally complex situations. Now, we will continue with a discussion of the first of these philosophical questions.

## 1.4 Meta-ethics and the problem with motivation

In this section, we take up the first philosophical question posed in §1.3, which concerns the meta-ethical assumptions that are necessary to formulate the problem with motivation for sustainable action. As stated earlier, Humanity's Challenge contains my assumptions about the main results of climate science, as well as the normative demand to collectively mitigate climate change. If, then, we want to study motivation for acting sustainably by considering agents who take moral action, it could be objected that Humanity's Challenge cannot motivate. This means that the motivational issue would need to be approached from a view that considers the agent's motivation as independent from the reasons that justify sustainable action (see §1.3). In this way, I will explain how the normative requirement to meet Humanity's Challenge can have motivating force on individual agents.

The internalism-externalism debate in meta-ethics (Rosati 2016; Gardiner 2011b; Brink 1997; Smith 1994) is about the connection between moral judgements and motivation – that is, it deals with the question: Do moral judgements motivate necessarily or do they motivate only contingently? In summary, internalism holds that an agent cannot sincerely make a moral judgement without being motivated to abide by her judgement. According to externalism, on the other hand, any connection that exists between moral judgement and motivation is purely contingent, though it may turn out to rest on deep features of human nature. As in the case of the homeowners (see §1.1), S is internally motivated if she derives her motivating reasons for installing the solar panels from the justifying reason as provided by Humanity's Challenge. She is externally motivated if she accepts the demand of Humanity's Challenge but is motivated to install the solar panels because of reasons that have nothing to do with Humanity's Challenge, for instance because she likes the colour of the panels more than the colour of her roof. Within the spectrum of internalist and externalist positions, I argue that a weak internalist view on motivation captures the possibilities both that agents feel, necessarily, *some* motivation to live up to Humanity's Challenge, and that this motivation can be overridden by conflicting desires (Rosati 2016, §3.2).

Humanity's Challenge may have limited motivating potential for individual agents for two reasons. First, the concepts of posterity and future generations seem too vague for current people to become concerned about the well-being of people who will live in the distant future. Humanity's Challenge may seem simply too abstract for current people to become motivated for sustainable action. More precisely, from a common-sense point of view, it seems undisputed that current people have specific moral obligations towards their children (Scheffler 2001, 36). This means that for current people, their interest in the well-being of future people starts to fade away not immediately – that is, during the lifetime of their children – but from some point in time thereafter. If we understand future persons as human beings with whom current people have no overlap in their lifetime, then it is safe to assume future generations to begin roughly one hundred years from now, after which the concern of current people for future generations diminishes rapidly (Ott 2004, 85–86). To sum up, although people feel motivated to care for their offspring, their concern for meeting Humanity's Challenge may be very limited.

Second, the gap between current and future generations tends to become wider because of what Samuel Scheffler has called the "phenomenology of agency". Under a common-sense understanding of human social relations "as consisting primarily in small-scale interactions with clearly demarcated lines of causation among independent individual agents", this is a way we experience ourselves as agents with causal powers (Scheffler 2001, 39). Common-sense morality[14] assigns to agents considerable discretion in the way they choose to lead their lives. Common-sense morality thus limits the impact of morality on an individual's life. Scheffler's phenomenology of agency consists of three elements.

First, it tells us that acts have primacy over omissions. Whereas we ordinarily experience our acts as acts, we experience our omissions as omissions only in specific contexts. In the context of mitigating climate change this is relevant, since many sustainable actions are based on our awareness that we omit to care about our luxury emissions. Second, near effects of our actions have primacy over remote effects. We tend to experience our causal influence as inversely related to spatial and temporal distance. Since future people, who will benefit from our sustainable actions, are temporally distant, this feature is relevant for our investigation as well. Third, individual effects have primacy over group effects. When an outcome is the result of the joint actions of a number of people, including ourselves, we tend to see the contribution of our own agency to a much lesser extent than we do when we take an effect to have resulted solely from our own actions. Since meeting Humanity's Challenge requires a concerted behaviour by many human beings and institutions, this feature is particularly important to consider (Scheffler 2001, 39).

We can see, therefore, how the interests of future people could become so remote from current people that Humanity's Challenge cannot motivate anymore. We might then be tempted to support a purely externalist stance on motivation for taking sustainable action. This would mean that Humanity's Challenge would not provide us with a motive to act sustainably, but, instead, our motivation for sustainable action would come about in a purely contingent way – that is, for reasons that are outside the challenge. As I argued in §1.3, the topic of this book would then become entirely empirical, since there would be no relationship between the normative reason to act on Humanity's Challenge, which we have accepted, and the motivating reasons that explain our sustainable action.

In response, I argue for two reasons that, for our investigation, a weak internalist position is a plausible assumption. First, current people have an interest in the well-being of future people: parents have an interest in the well-being of their children, and current people in general have an interest in the existence and well-being of future people, simply because they would like to receive healthcare from future people when they are old. Since current people cannot reasonably deny others coming after them the same level of care they receive, they will judge it morally good that at least a couple of the future generations come into existence. Accepting Humanity's Challenge implies accepting the moral judgement that we ought to mitigate climate change by keeping the global temperature rise below 2°C. Since future people need a global temperature rise of not more than 2°C for their well-being, this judgement connects the agent's interest, in the form of the well-being of her offspring, to the interest of humanity.

Current people want the next generation to grow up in a world where climate change is under control, and the next generation wants the same for the generation thereafter, and so on. Each generation has an interest in the well-being of the next generation in the sense of mitigated climate change. This means that, although the agent may not consider Humanity's Challenge as a motivating reason, and regardless of how small the agent considers her contribution to the problem of climate change, she has at least *a* moral motive to act in line with what the challenge commands (Birnbacher 2009, 286).

Second, adopting an externalist view on motivation implies that agents would not be able to keep track of changes in their moral judgement and corresponding changes in their motivation. Michael Smith argues for this claim for "the good and strong-willed person" (1994, 71–76). For example, because of the actual climate situation the people of the islands of Tuvalu are anticipating the loss of their territory rather sooner than in the far future (Farbotko and Lazrus 2012). This could, in turn, increase the urgency of Humanity's Challenge and change the motivation of individual agents in affluent countries. It thus seems implausible that an agent's acceptance of the challenge would not lead to any motive for sustainable action whatsoever.

For the above reasons, I consider weak internalism to be an adequate assumption for investigating the problem with motivation for sustainable action. This allows us to presuppose in this book that agents feel a pre-cognitive motivation to live up to Humanity's Challenge, and also that this motivation can be overridden by conflicting desires (Rosati 2016, §3.2). Like Birnbacher (2009, 282),[15] we are focused on the barriers to tapping into the sources of motivation that are required *over and above being convinced* by the normative reasons in favour of collectively meeting Humanity's Challenge. Our investigation will, in Chapter 9, lead to the claim that, in addition to the moral motives for sustainable action, other "indirect" motivations are needed to act on the challenge.

The acceptance of my argument for an assumption of weak internalism, together with my claim that there is evidence that the problem with motivation for sustainable action is distinctive, since there are four philosophical questions to be dealt with that may not be relevant in the same way in other morally complex situations (see §1.3), allows us to continue with analysing the effect of Humanity's Challenge on the motivation of individual and institutional agents. Irrespective of the continuing internalism-externalism debate, we have at least *a* motive to take up Humanity's Challenge and start acting sustainably.

## 1.5 The problem with motivation for sustainable action

We will now formulate the problem with motivation for sustainable action that we will investigate. To remind us that we are still working on an actual problem, let us consider a statement from the latest IPCC report:

> Pathways limiting global warming to 1.5°C with no or limited overshoot would require rapid and far-reaching transitions in energy, land, urban and infrastructure (including transport and buildings), and industrial systems (high confidence). These systems transitions are unprecedented in terms of scale, but not necessarily in terms of speed, and imply deep emissions reductions in all sectors, a wide portfolio of mitigation options and a significant upscaling of investments in those options (medium confidence).
>
> (IPCC 2018, 17)

Therefore, as we have assumed, Humanity's Challenge remains an actual and urgent challenge that we, the current generation, must overcome. The current generation is uniquely placed in human history:

> The choices we make now – in the next 10–20 years – will alter the destiny of our species … unalterably, and forever. Generations prior to us did not know the havoc they were wreaking, and it will be too late for those who come after us.
>
> (McKinnon 2012, 129)

Given the urgency of the situation, it is surprising that so little attention has been paid to the issue of motivation. To the best of my knowledge, there are just a handful of *philosophers* who have published works about the problem with motivation for sustainable action in the sense that they have discussed the problem head-on as the *main topic* of their work.[16]

Since human agents are capable of acting efficaciously in many other circumstances, it seems to me that there are differences in our motivation to act (where I understand 'act' to include intentionally refraining from acting) between situations where a sustainable action is required and other morally non-trivial situations. As Humanity's Challenge has been formulated, I understand our individual and collective failure to take up the Challenge to be due to a lack of motivation for taking sustainable action. As I see it, this lack of motivation can be considered as the interaction of three fundamentally different problems, namely those of individual agents and their motivation to act sustainably; institutions whose policies must enable taking sustainable action; and humanity as a collective of human beings that ought to understand its own existence in relation to the planet. Let me explain these three problems.

First, as *individual* agents, especially in affluent countries, we seem in some way seriously hampered from making the step from what we judge *worth doing collectively* to actually *doing* it. The problem with motivation for sustainable action for individual human agents appears as the problem of overcoming the inconsistency between the *moral claim* in favour of collective environmentally friendly action, and the *action* that we in fact carry out as an individual agent following up on that norm. Note that in psychology this is generally called the "attitude-behaviour gap" (Gifford 2011, 290).

We started this chapter with the example of a group of homeowners who react to a proposal for a sustainable action by asking what they will get in return for taking such an action ("What's in it for me?"). The verbal and non-verbal reactions of the other homeowners tell us that they consider S's proposal to buy nine solar panels for each home somehow as unconventional. More precisely, we considered the scenario of a meeting held in a corner of some middle-class town in an affluent country, where climate science was endorsed, and where the norm was agreed that collectively the homeowners ought to reduce their greenhouse gas emissions. In this situation, it came as an unexpected, unpleasant surprise to the other homeowners that S proposed to commit to the supplier's offer, whereas in S's view this was the natural thing to do after accepting the norm to strive for a climate neutral neighbourhood. S's peers seemed to consider it unproblematic not to be moved by Humanity's Challenge to promote sustainability. In questions concerning the requirement to take sustainable action, to which we have agreed as a norm for humanity, we thus seem to be inclined to make our commitment dependent on what we will get in return for our efforts. Let us look at two other problem formulations.

Christoph Baumgartner discusses the motivational "deficit" that remains between accepting a norm to contribute to a more sustainable world on the one hand and acting according to that norm on the other (Baumgartner 2005, 34). Baumgartner makes two restrictions. First, he focuses on the daily actions related to the environment by individual actors in Western industrialised countries. Second, he confines the daily actions of human agents to actions in their role as private individual only. Typical examples of actions taken into consideration include: the choice of means of transportation (bicycle, car, aeroplane, etc.), buying environmentally friendly products in the supermarket for private consumption, and saving energy and water in the private household. This means that Baumgartner's account needs to be supplemented by a similar study of actions by persons in their role as professionals in the field of business and in other organisations (Baumgartner 2005, 27–28).

Similar to Baumgartner, Dieter Birnbacher formulates the problem with motivation of individual agents as bridging the gap between the acceptance of a norm and the action that follows up on it (see §1.3). Since sustainable action cannot rely on moral motives only, Birnbacher looks for other indirect motivations that "do not aim at the production of goods or the prevention of evils befalling future people but aim at objectives in the present or in the near future" (Birnbacher 2009, 285).[17]

Second, the problem of not acting sustainably appears as a problem for *institutions*. The ongoing United Nations climate negotiations that started with the UNFCCC in Rio de Janeiro (1992) and led, most recently, to the climate summit in Paris (2015) indicate that sovereign states at least have made an increasing commitment to making sustainable development happen. However, given the above-mentioned scientific judgement that global greenhouse gas emissions are still increasing and have accelerated over recent decades, it would be overly optimistic to believe that the latest climate negotiations have made the motivational problem at the institutional level redundant.

For example, in the Netherlands, in a historically unique lawsuit, an independent judge has ruled that the state has been negligent in keeping its promises regarding the mitigation of climate change. Amongst other things, the court ruled that

> the State must do more to reduce the emission of greenhouse gases in the Netherlands. The State must ensure that the emissions in the Netherlands in 2020 are at least 25% lower than in 1990. … Based on current policies of the State, the Netherlands will reach a reduction in 2020 of 17% at the most. This is below the norm of 25 to 40%, which climate science and the international climate policy deem necessary for industrialised countries.
>
> (*Urgenda* 2015)

This lawsuit highlights that politicians as the agents of their voters can be held to account for not delivering those institutional policies that are necessary to realise the promises they have made to mitigate climate change. Hence, even if we were able to fully understand the motivational drivers for sustainable action at the individual level, at the level of institutions many questions about the problem of motivation remain.

The problem of sustainable action at the institutional level requires high levels of social coordination and intersubjectively shared values and intentions by the agents involved. In liberal democratic societies, for a policy on sustainable action to be approved by the political majority, it must be understood and supported by the public at large. This means that promoting a policy on sustainable action, even if this were only to limit one's freedom to cause "luxury" emissions, requires an understanding of the motivational problems of individual agents, as otherwise it will not be supported. Thus, although the problem of sustainable action at the institutional level is of a different order, it seems strongly connected to the problem at the level of the individual agent. Other formulations of the institutional problem with motivation for sustainable action are given in the accounts of Care and McKinnon.

Norman Care emphasises the connection between the problems with motivation at the individual and at the institutional level when he places the reader in the position of an independent policymaker in a free society, who takes on the task to "formulate policies and put them before the larger community for approval" (Care 1982, 195). Care proposes a policy that conserves our society's legacy for future generations, whilst he acknowledges that such a policy might require ordinary people to sacrifice a serious part of their wealth. This is how Care formulates the problem:

> What (apart from legal or other forms of coercion) is available in the way of support for the morally principled policies we now see we must recommend to the larger community? Alternatively put: what in people can we draw upon to *motivate* them to follow policies that morality requires regarding our legacy for the world of the future?
>
> (Care 1982, 197, emphasis Care)

In this version of the motivation problem, Care thus argues that an institutional policymaker must understand and draw upon what motivates the individual agent to take sustainable action, since in a liberal democracy such a policy will otherwise not be approved.

McKinnon identifies the following as the key problem for a politics of climate change: "How can we motivate the rich to accept, indeed to initiate, the changes to their lifestyles that are required for bottom line greenhouse gas emissions reductions?" (McKinnon 2011, 197). Like Birnbacher, McKinnon notes that the motivating force of an appeal to an impartial principle of justice seems already questionable in the intragenerational case, let alone in the intergenerational case, "given that we will never see the suffering caused to those by our unjust actions, and never run the risk of coming face to face with their pain" (McKinnon 2011, 198).

Third, the universal problem for *humanity* that hampers sustainable action is how we practically understand ourselves as human beings on a finite planet against the background of taking up Humanity's Challenge. To better understand what this problem is about, we should discuss the view of Stephen Gardiner, who has pointed at the concept of moral corruption as the root cause of an agent's attitudes of complacency and procrastination when it comes to taking environmentally friendly action.

According to Gardiner, the implication of an overemphasis on the global complexity of the problem of climate change is that the intergenerational argument for taking sustainable action does not receive attention, since this "may turn out to be *perfectly convenient* for us, the current generation, and indeed for each successor generation as it comes to occupy our position" (Gardiner 2006, 408, emphasis Gardiner). The way we choose the terms in which we discuss the problem of climate change may be influenced by our wish to maintain our convenient status quo. This exploitation of our global and intergenerational position may be called the problem of moral corruption (Gardiner 2011a, 8). Gardiner's main thesis is that we do not act sustainably because of our vulnerability to moral corruption (Gardiner 2011a, 22). With this concept, Gardiner intends to address the fact that we as agents in affluent countries are not only in a position to pass the buck to future generations and the poor, but that we also face strong temptations to do so.

Above, I have formulated the problem addressed by this book as the blend of three problems that mutually influence each other: individual agents and their motivation to act sustainably, institutions whose policies must promote sustainable action, and our practical self-understanding as a collective of human beings that ought to include the imperative of a sustainable future for the planet. I will now discuss the moral community of agents that ought to take up Humanity's Challenge.

From the first UN climate summit in Rio de Janeiro (1992) onwards, it has been acknowledged that the developed countries should take the lead in mitigating climate change. These countries, as listed in Annex I of UNFCCC (1992), "shall adopt national policies and take corresponding measures on the mitigation of climate change, by limiting anthropogenic emissions of greenhouse gases and protecting and enhancing its greenhouse gas sinks and reservoirs" (UNFCCC 1992, Art. 4.2 (a)). Furthermore, within this group of countries there is a subgroup ("Annex II countries"), which has committed itself to provide financial resources to the developing countries for them to meet their obligations under the agreement, and to assist them in meeting the costs of adaptation to the adverse effects of climate change (UNFCCC 1992, Art. 4.3–4.4).

In other words, from the start of the climate change negotiations it seems that the UN member states have acknowledged that to certain developed countries a "can-implies-ought" (Garvey 2008) argument applies, as laid down in their Annex II status. Approximately one billion people inhabit this group of countries. The members of this group are all constitutional democracies; they all belong to the "high income OECD countries" (i.e. having an on average Gross National Income per capita of $44,000 (World Bank 2018)); and they all have a very high score on the Human Development Index (i.e. 0.8 or higher), which indicates that their inhabitants in relative terms live a long and healthy life, are well educated, and have a decent standard of living (United Nations Development Programme, 2018). Because of the special moral status of the group of Annex II countries in the debate about who ought to take action in mitigating climate change, we confine the moral community that is addressed in this book to this group of countries. The terms "Annex II countries", "high-income OECD countries", "affluent countries", and "highly industrialised countries" will be used interchangeably.[18] The project of meeting Humanity's Challenge thus importantly relies on the possibilities and motivation of individual and institutional agents in the affluent countries for taking sustainable action.

Alternatively put, the countries in the scope of this book belong to the group that Elizabeth Cripps describes as "The Able: The potential collectivity composed of those (the global affluent) who could contribute to action on climate change mitigation and adaptation at less than significant cost to themselves" (Cripps 2013, 203).

As I see it, for the project of mitigating climate change to succeed, it follows from a weak version of "ought-implies-can" (Stern 2004, 46) that ethicists should go beyond debating the normative requirement for mitigating climate change only. Since it is now clear that having moral motives only is not sufficient for acting sustainably, as ethicists we are also interested in the obstacles that impede an agent's motivation. Presupposing that we have accepted Humanity's Challenge as a norm, we go on to investigate what it means for agents to live up to this challenge, given that their options to act sustainably may be limited and the (motivational) force – as psychologists of moral motivation call it – that pushes them to take sustainable action may be weak (Heinrichs *et al.* 2013, 1). Within this space of possibilities for taking sustainable action, the hindrances that agents have to cope with are of special interest to us.

## Notes

1 Ott and Döring (2008, 22–34) provide an overview of the history of sustainability and sustainable development.

2 In 1994 this obligation was implemented as part of the German constitution, of which Article 20a reads:

> Mindful also of its responsibility toward future generations, the state shall protect the natural foundations of life and animals by legislation and, in accordance with law and justice, by executive and judicial action, all within the framework of the constitutional order.

3 Birnbacher (2015, 156) also refers to the rebound effect to emphasise that "future-oriented norms, laudable as they are, largely fail to make an impression on the motives of individual and political agents".

4 Think of the inhabitants of the islands of Tuvalu in the Pacific, who are considered among the first climate refugees (Farbotko and Lazrus, 2012), whereas it is mainly countries in Asia, Europe, and Northern America that are responsible for historical $CO_2$ emissions (Olivier *et al.*, 2012).

5 Gardiner (2011a, chapters 7–8) discusses what he calls the "theoretical storm". In addition to economic cost-benefit analysis, Gardiner points at the failure of political institutions and political theory to deal effectively with the consequences of climate change.

6 Elizabeth Cripps makes a similar distinction between climate science and normative ethics (2013, 4).

7 Note that this study has been criticised by Richard Tol (2016) for overestimating the consensus among climate scientists. In response, Cook *et al.* (2016) criticise Tol for using a flawed methodology. They reaffirm that "the consensus that humans are causing global warming is shared by 90%–100% of publishing climate scientists" according to six independent studies by the co-authors of their paper.

8 The view in philosophy on the two groups of theories is supported by a recent meta-study on the psychology of moral motivation, covering 97 papers, grouped around the question: "How Can Moral Psychology Save the Planet?" The authors conclude that moral psychology can facilitate moral responses to environmental issues, if an appeal is made to harm-based theories (current and future generations, animals) or purity-based theories (ascribing value to the environment, which is worthy of being protected at all costs) (Rottman *et al.*, 2015).

9 See for an introduction and main topics: Gosseries and Meyer (2009); for a systematic account: Unnerstall (1999); and for more on the recent debate: Bos and Düwell (2016).

10 Unnerstall (1999) and Ott (2004) have studied the main ethical theories (utilitarianism, Rawls, discourse ethics, categorical imperative) and their suitability for constituting a basis for the foundation of X. Unfortunately, a discussion of the various options for Y falls outside the scope of this book.

11 Assessing the normative implications of the concept of natural capital becomes complex. For example, as Ott and Döring argue, we should be aware that applying a very broad definition of natural capital could lead to a kind of magic word, and the possibility of reasoning in practical syllogisms taking place as follows: (1) It is necessary to protect natural capital; (2) X is part natural capital; (3) hence, X must be protected (prima facie). Indeed, this would leave agents with few options to act, as their actions will very often affect *some* part of nature. This means that, in turn, the normative implications lead to the choice of a more restricted definition of natural capital (2008, 223). Again, this task falls outside of the scope of this book.

12 Birnbacher (2015, 154) comments: "The more general challenge is to achieve sustainability in man's dealings with nature in a world with continuing population growth and rising expectations of material well-being."

13 Of course, if we consider the motivation of individual human agents, there are numerous *psychological* questions as well with regard to being motivated to take sustainable action. See for various topics in environmental psychology: Bechtel and Churchman (2002), Kollmus and Agyeman (2002), Schmuck and Schultz (2002), Vlek and Steg (2007), Steg and Vlek (2009), Koger and Winter (2010), Steg *et al.* (2013).

14 Common-sense morality is construed as the default moral position of agents, which they generally tend not to analyse and discuss. It allows agents to give weight to their personal interests with regard to ethical theories by establishing a personal point of view as opposed to the moral point of view (see §4.3).

15 Similarly, Peeters *et al.* argue for weak internalism that provides an alternative to externalism as well as preventing a full revision of already developed moral concepts (2015, 44).

16 The main publications are: Care (1982), Baumgartner (2005), Birnbacher (2009), McKinnon (2011), Bourban (2014), Peeters *et al.* (2015), and Coeckelberg (2016).

17 Following Birnbacher, Michel Bourban (2014) looks for quasi and non-moral motives that might be able to motivate for sustainable action.

18 At all times these groups of countries are to be taken to consist of: Australia, Austria, Belgium, Canada, Denmark, European Union, Finland, France, Germany, Greece, Iceland, Ireland, Italy, Japan, Luxembourg, Netherlands, New Zealand, Norway, Portugal, Spain, Sweden, Switzerland, United Kingdom, United States of America.

# 2 Methodological considerations

## 2.1 Introduction

In this chapter, we continue with the groundwork by taking two steps. First, we take up the question: What or who should be considered the addressee of the moral requirement to take sustainable action? It only makes sense to continue with studying the motivation problem if we have found an argument that individual human agents in affluent countries share *some* responsibility for making sustainable development happen. To make the point, I will outline views that argue in favour of downplaying the responsibilities of individuals. Second, the method of investigation of the book will be outlined. I will answer the question: How can I as a philosopher say something relevant about the motivation of agents? In response, I will argue for a conceptual-phenomenological analysis of the first-person point of view of the agent that potentially can be aligned with the methods of empirical psychology. These topics will be dealt with in separate sections of the chapter, as now outlined.

The question about the addressee of the moral requirement to take sustainable action must first be answered (see §1.3), since it otherwise becomes entirely unclear who or what is supposed to act on Humanity's Challenge. As already stated, this question becomes relevant since I have argued in favour of individual agents having a perhaps rather weak motive for taking up Humanity's Challenge (see §§1.3–1.4). In response, I argue that not only political institutions but also individual agents in the affluent countries bear some responsibility for meeting Humanity's Challenge, regardless of what the institutional policies require (§2.2).

Then, in order to address the way of investigating an agent's motivation in a second step, a methodological approach needs to be developed that meets two criteria. First, the method should allow for a connection to philosophy in the sense that the observations of agentive motivation can be integrated into a philosophical theory of motivation for specifically sustainable action. And, second, as the concept of motivation (see §1.4) at first sight seems to have many psychological ramifications, it should be able to relate to methods of empirical psychology, such that the observations of agentive motivation can be understood and categorised by mainstream psychological theories. To satisfy these two criteria, we start with an orientation in two psychological disciplines: the psychology of moral motivation and environmental psychology. Both disciplines have traditions that enable the method of this investigation to be empirically informed (§2.3).

In §2.4, we continue with an explanation of the method of this book, which takes an understanding of the first-person perspective of human agents as its starting point. As it focuses on the first-person perspective, the investigation can be understood as phenomenological. I will outline an approach to phenomenology that stays close to a narrow interpretation of this concept and that remains consistent with mainstream positions in the philosophy of mind. Finally, I will qualify the method as descriptively diagnostic. This method of understanding the first-person perspective leads to emphasising the pre-empirical nature of the findings of this book. That is, the descriptive diagnosis developed in the book aims to propose hypotheses for further empirical research.

If we want to compare the motivational features of action situations where sustainability is at stake, we need an empirical concept of human action that enables us to describe and compare these actions in a sufficiently fine-grained manner. The empirical Rubicon model of action phases seems a candidate for performing this role. It divides a course of action into four consecutive, distinct segments or phases: deliberation, commitment, action, and evaluation. It starts with a person's deliberation about her options for action and it ends with the evaluation of the achieved outcome of the action. The phenomena of committing to the chosen goal and executing the required actions are situated in between. The Rubicon model thus also serves the purpose of breaking down the question "Why is it hard to engage in sustainable action?" into specific questions that relate to the action phases of the model (§2.5).

## 2.2 Collective, institutional, and individual responsibility

In §1.3 we posed the question what or who should be considered the addressee of the moral duty to perform sustainable action. This question will often have an obvious answer when concrete action situations are considered, where specific agents are under pressure to execute strategies that should lead to solving specific problems. In the case of climate change, however, it seems unclear how the responsibilities for taking the mitigating actions should be allocated between institutions (e.g. states) and individuals.

In this section, I will argue that, from the collective responsibility at the level of sovereign states, for example as assumed in the UNFCCC, it follows that, even if the political institutions that signed up to the framework have primary responsibility, most people in highly industrialised countries have at least *some* moral responsibility to act in accordance with Humanity's Challenge. In my argument, I need to take a view on the concept of institutions, which will form an assumption for the remainder of the book. I make the assumption that institutions consist, essentially, of human individuals only. However, as I will explain, this does not mean that the responsibilities of institutions can be reduced to the responsibilities of individuals.

In the debate about what or who should be the addressee of the requirement to implement sustainable development it has been argued that the responsibility for implementation should be left to institutions such as governments and corporations. Another view is that realising such development is a matter that cannot be left to individuals, since it is irreducibly collective (Woods 2015, 97). Walter Sinnott-Armstrong looks for a moral principle on which it can be argued that the facts about global warming associated with wasteful driving (see §1.2) impose upon us a moral obligation not to drive a gas-guzzler just for fun on a sunny Sunday afternoon (Sinnott-Armstrong 2010, 334). With the example of wasteful driving in hand, Sinnott-Armstrong tests no fewer than fifteen moral principles, ranging from a harm principle to a contractualist principle. Sinnott-Armstrong concludes that none of these principles is capable of imposing upon us the moral obligation not to go for the drive, in order to claim, finally, that "global warming is such a large problem that it is not individuals who cause it or who need to fix it. Instead, governments need to fix it, and quickly" (Sinnott-Armstrong 2010, 344). This brings him into the camp of those who emphasise that it is the institution rather than the individual agent that has the moral obligation to do something about mitigating climate change.

In order to formulate my account, I do not need to discuss here the principles Sinnott-Armstrong reviews, and whether they meet the test of creating for us a moral obligation not to go for the drive on a Sunday afternoon.[1] Instead, my problem with Sinnott-Armstrong's argument lies in a presupposition that he seems to take as self-evident. Sinnott-Armstrong considers the drive on a sunny Sunday afternoon so insignificant regarding its greenhouse gas emissions that it is neither necessary nor sufficient for global warming (Sinnott-Armstrong 2010, 334). In other words, the aggregate amount of greenhouse gas emissions, due to the driving by human beings, causes climate change, whereas each individual drive does not. Since this assumption holds, in Sinnott-Armstrong's view, the agent is discharged of an individual moral obligation to refrain from the wasteful drive.

This is what Avram Hiller criticises when he asks which "odd emergent entity that is not reducible to individual acts of driving" would then be the cause of climate change (Hiller 2011, 354). Human beings are discrete, and their number is finite. The greenhouse gases emitted by our individual drives thus simply add up to having an aggregate effect on climate change. Hence, although at the level of global greenhouse gas emissions the impact of my individual car ride may be insignificant, it is by no means zero. As the drive is an action that is done purposively and voluntarily, the agent could have reduced her contribution to climate change by not taking the drive. In a similar vein, Christopher Morgan-Knapp and Charles Goodman argue that under act-consequentialism the arguments are mistaken that the effect of individual emissions are minuscule, and that vanishingly small impacts are morally irrelevant because no one could possibly tell the difference between such an impact occurring and it not occurring (Morgan-Knapp and Goodman 2014).[2] Hence, it seems that one cannot ignore the effect of one's individual emissions on global warming. The question that follows for people in the affluent countries is whether they are under a moral obligation as individuals to take action against global warming. Sinnott-Armstrong, however, assigns the obligation to act against climate change exclusively to governments, which he assumes to be capable of taking the appropriate measures (Sinnott-Armstrong 2010, 332).

Jessica Nihlén Fahlquist discusses the boundaries between individual and institutional moral responsibility for environmental problems. In line with the argument for assigning the task of taking up Humanity's Challenge to the affluent countries (see §1.5), she argues that individuals in these countries should have ascribed to them a forward-looking responsibility for mitigating climate change (Nihlén Fahlquist 2009, 118). That is, individuals in the affluent countries have the responsibility to take some action so as to prevent climate change. However, Nihlén Fahlquist also emphasises the structural and societal features that work against individuals actually taking sustainable action. These obstacles have to be taken away by governments and corporations, "because they can create systems that make it easier and less costly for people to choose the environmentally friendly option than to choose the environmentally harmful option". Institutions, in the view of Nihlén Fahlquist, have the task and the power to implement policies that are efficient and fair, since they apply to large groups of individual agents, enabling them to take sustainable action (Nihlén Fahlquist 2009, 120).

Regarding Nihlén Fahlquist's account, it seems to me that her metaphysics of social institutions could be a lot clearer. On the one hand, she considers it the duty of individuals to change these institutions such that the institutions can take away obstacles for sustainable action. On the other hand, she clearly discriminates between the responsibility assigned to individuals and the responsibility assigned to so-called institutional agents. Finally, Nihlén Fahlquist nominates the human individual as the addressee of the requirement to take sustainable action, whereas in her argument she treats institutional agents as if these are separate ontological entities. Indeed, it is my view that both Sinnott-Armstrong and Nihlén Fahlquist should have made explicit at the outset their presuppositions about the social cooperation of human beings in an institutional context.[3]

While I agree with Nihlén Fahlquist that the political institutions have their own responsibility for taking up Humanity's Challenge, I do not agree with Sinnott-Armstrong that these institutions alone can do the job and that individual agents are under no obligation to end their non-sustainable behaviour (see also Gardiner 2011b, 52). I also agree with Nihlén Fahlquist, in following Henry Shue, that it is an indirect duty of individual agents to make sure that there are adequate institutions, whose task it is to strive to meet Humanity's Challenge.[4] The point on which I criticise both authors is that in their accounts the realms of individual and institutional agents tend to become normatively separated. This means that a person who occupies both individual and institutional roles may be tempted to morally evaluate her agency while using incompatible sets of normative standards. Consequently, the responses by Nihlén Fahlquist and Sinnott-Armstrong to the question what or who should be considered the addressee of the requirement to take sustainable action remain abstract. To clarify the position of an agent, who wants to consider her various roles from an understanding of herself as someone who operates from an integral set of normative standards, we need to make some assumptions about the concept of institutions.

While remaining true to the extensive literature on the concept of institutions (see e.g. Gledhill 2014; Miller 2012; Murphy 1999; Searle 2005; Shue 1988), I want to emphasise that social institutions consist of human persons, who occupy institutional roles (Miller 2012, §5). That is, the institutional role occupants have to integrate their role within an understanding of themselves as practical human beings. Such integration is to be understood in the above sense of being able to morally evaluate one's different roles as against an integral set of normative standards. What agents do in their role as institutional agent and what they do as private individual, or in another institutional role, have to be integrated within their overall conception of themselves as a person (Miller 2012, §1). Since it is obvious that taking up Humanity's Challenge requires institutional policies that will affect many people, while they occupy individual and institutional roles, we need a concept of institutions that allows people to reconcile their various roles with the overall understanding of themselves as a person. Thus, a requirement for the concept of institutions is that it enables the *institutional* role occupant to understand what it means to be affected by a prospective policy while she is performing her *individual* roles. This understanding is necessary to ensure that the policy will be effective.

For the debate about what or who should be considered the addressee of the requirement to perform sustainable action, I therefore take as a methodological assumption a minimally individualistic concept of institutions (Schmitt 2003, 7). That is, I assume that organisations and institutions are comprised of human individuals only. Institutions are composed of institutional agents, who are in an ongoing discourse with each other about the strategies, policies, and actions they are pursuing as representatives in view of the goal of the institution. The conception of what the institution entails is in fact intersubjectively shared among the representatives, who may perform different tasks involving different responsibilities. The institution is upheld by the ongoing discourse between the representatives. All actions that in common-sense language are ascribed to the institution are in fact carried out by representatives, who act according to a commonly shared role description, or at the very least a personal interpretation of such a description, for which the representative can be held to account in the light of the purpose of the institution.

Understanding institutions in this way, however, does not imply that the responsibilities of institutions can simply be reduced to individual responsibilities. In the words of John Searle, a social institution should be seen as a "fictitious person", meaning that the institution "is not identical with any physical object or any person or set of persons" (Searle 2005, 17). Hence, social institutions are entities that can be held accountable retrospectively and can assume prospective responsibility. They are capable of entering into contractual relationships, buying and selling, making a profit, and incurring debts, for which they are liable. However, actions taken by the institution always occur through human agents, who represent the institution in responding to attempts to call the institution to account. The prospective responsibility of institutions also works through human individuals. Statements such as "Institutions must contribute to the mitigation of climate change", in my view, should be interpreted as follows: Those representatives of the institution who are supposed to take up the task of mitigating climate change are the ones who are responsible for taking the required action. If it is unclear who is supposed to do this, then it is up to the top management of the institution to ensure that the task is delegated to those representatives who are capable of taking up this role.

I have already mentioned that the collective part of the problem with motivation for sustainable action can be illustrated by considering the UNFCCC framework (including its successor, the Paris Agreement of 2015) as a commonly shared normative requirement being embodied in an agreement between 195 UN member states. The UNFCCC has now been in place for more than two decades, and we have included its essence in Humanity's Challenge. As we have seen in §1.5, the science of climate change indicates that the realisation to date of the goals laid down in the UNFCCC is far from where it should be. Given the lack of major progress in realising the objectives of the framework since 1992, it could be argued that the agents responsible for taking the necessary actions (e.g. those politicians who signed the framework) face motivational problems in following up on the framework's moral claims, and, in any event, could be subject to strong ethical criticism (Gardiner 2011a, 404). However, the lack of progress in living up to the UNFCCC is a moral problem not only for those who are responsible in their role as politicians, but also for many of us in the affluent countries.[5] My argument for this claim is as follows.

First, we have seen that we should presuppose that scientists are right in their judgement that current greenhouse gas emissions are too high to keep global warming within the level aimed for by the UNFCCC (see §1.3). I then argued that theories of intergenerational justice justify the norm that, as a collective, present human beings have an obligation to prevent a global temperature rise of more than 2 °C (see §1.3). On the basis of the scientific projections, this means that any responsible policy would require that, at some moment in the not too distant future, global emissions of greenhouses gases should peak and from there on start decreasing substantially for at least a couple of decades. This global course of action to reduce greenhouse gas emissions must continue until the desired levels are reached. As the IPCC puts it:

> There are multiple pathways that are *likely* to limit warming below 2°C relative to pre-industrial levels. These pathways would require substantial emissions reductions over the next few decades and near zero emissions of $CO_2$ and other long-lived greenhouse gases by the end of the century.
> (IPCC 2014b, 15, emphasis IPCC)

If we add to this the fact that the least advantaged countries have the human right to increase their "subsistence emissions" (Shue 2010) to achieve a certain level of wealth for their inhabitants (Caney 2010, 172), it follows that the affluent countries must reduce drastically and start to do so immediately.

Second, if my first point is true, and the rough task of mitigating climate change is clear,[6] then who is responsible for making it happen? Following my minimally individualistic understanding of the concept of institutions, it could be argued that the responsibility for taking up Humanity's Challenge ought to remain with the fictitious person of humanity. In other words, Humanity's Challenge is a challenge only for humanity. Humanity, understood as the collective of human beings presently populating earth, would then be accountable for climate change; however, no other role occupant would be under the duty to take sustainable action. We should note at this point that, in addition to the roles of individual and institutional agent, human beings occupy a third role that they must evaluate from an overall understanding of themselves as a person. This is the role of being a *member* of humanity. Being human means that one is part of humanity. Since we must consider ourselves members of humanity, integrating this role into understanding ourselves as a person means that, as individual agents, we cannot disregard the obligation of humanity to safeguard its own existence through meeting Humanity's Challenge.

Given our task to morally evaluate our different roles as against an integral set of normative standards, it follows that the duty to mitigate climate change, which we each face in our role as a member of humanity, also affects our roles as individual and institutional agents. It would be normatively incoherent if we were to accept that we must do something about mitigating climate change in our role as humanity's member, while simultaneously disregarding this duty in performing the other roles. This means that we need to answer the question how humanity must be organised in order to take up the responsibility to meet Humanity's Challenge.

Again, we shall stay close to the roles of individual and institutional agents that human beings occupy next to their role as a member of humanity. Organising humanity such that it can take up Humanity's Challenge, then, must be compatible with one's roles as individual and institutional agent. Therefore, as human beings we must contribute to taking up Humanity's

Challenge not only as a member of humanity, but also as an individual agent or in our role as institutional agent.

Taking another step in the argument for individual responsibility for sustainable action, we can make use of the characteristics of the countries that comprise the moral community that ought to take up Humanity's Challenge (see §1.5). When looking for competent institutions that must work on this task, it could be argued that institutions in the highly industrialised countries are not suited to handle problems of the magnitude and complexity of the global climate change problem (Held 2014). Even national states could be seen as incapable of designing and implementing policies that promote sustainable action. Recall that we have assumed that these institutions are constituted by institutional agents who are in an ongoing discourse with each other about the strategies, policies, and actions they are pursuing in view of the goal of the institution. This means that, although the institution remains responsible for taking up Humanity's Challenge, its incompetence to come up with the appropriate policies and explain them to the public follows from the network of institutional agents not being capable.

Because the affluent countries are constitutional democracies, however, a process of delegation of individual responsibilities and powers has helped create many of these institutions, on the principle that the problems of each and every individual can be solved efficiently and effectively when the institution can organise the implementation of their solution (Gardiner 2011a, 432). In other words, in the high-income OECD countries there is a principal-agent relationship between voters and politicians, respectively. This implies that in the case of politicians being given the authority to propose and implement plans for reducing greenhouse gas emissions, and failing to do so, then these politicians can be held accountable and can be the subject of moral criticism. On the other hand, following my minimally individualistic understanding of the concept of institutions, the delegation of responsibility and powers from the individual voters to the institutions does not discharge the individual agents fully of their responsibilities in the case of the politicians failing to do their job. With or without responsible politicians, the problem of acting on Humanity's Challenge still remains, meaning that if delegation to the agents does not work, then the responsibility to act sustainably falls back on the principal – that is, the individual citizens. Hence, in the event that delegation to politicians fails, the individual citizens are likewise subject to moral criticism. This means that citizens in high-income OECD countries cannot entirely waive their responsibility for failing to achieve the goal of meeting Humanity's Challenge.

To conclude: people in the affluent countries, on my analysis, can only evaluate their roles as individual and institutional agents, and their role as a member of humanity, as against an integral set of normative standards, if they assume *some* individual responsibility for taking up Humanity's Challenge.

However, even if we accept individual responsibility for people in the affluent countries, it could be argued that doing something about the motivational problem in taking up Humanity's Challenge is simply unfeasible. That is, we might be tempted to believe that individual agents will not be able to become motivated to live up to the responsibility that has been assigned to them for taking sustainable action. What might remedy the problem instead is to create new mechanisms and technological fixes in high-income OECD countries, to steer our behaviour towards more sustainable action. The psychology of nudging could serve as an example in this regard. Citizens of a particular country, for example, could be gently stimulated ("nudged") to decrease their energy consumption by being shown – in an anonymised way – the annual energy bill of their neighbours (Thaler and Sunstein 2008). Generally, this will indeed stimulate the lowering of energy usage, but as a result of psychological manipulation, not moral reasoning.

For three reasons, however, there are problems with the view that institutionalised mechanisms and technological fixes should be preferred to solve the motivational problem for individual agents. First, choosing to create new mechanisms and technological fixes, or change existing ones, in order to solve the motivational problem of individual agents, following my concept of institutions, presupposes that some human individuals are motivated to carry out these tasks. In order to accomplish the task of creating new institutions or changing existing ones, the agents leading such institutions have to be convinced and motivated to take sustainable action as institutional agents. This means that they not only have to take sustainability seriously at home or in other roles in the private sphere, but also must commit to implementing institutional policies that stimulate sustainable behaviour. Given this, we have to understand what it means to be motivated as an individual agent to take up Humanity's Challenge.

Second, as Norman Care has argued (see §1.5), and given that the scope of this book is the highly industrialised countries, in order for an independent policymaker to propose a policy to those whose behaviour needs to change, she needs to understand the motivational problems of individual agents. The policymaker should be able to explain to the affected people, who will approve the policy in a constitutional democracy, why this policy will most likely have the desired effect of promoting sustainable action. To the people who will approve the policy it must be explained why it is worth paying the price of changing their behaviour by potentially limiting their freedom to produce "luxury" emissions, and accepting the institutional force to achieve a more modest life as far as greenhouse gas emissions are concerned. For this reason too, then, we have to understand what it means for individual agents to be motivated to take sustainable action. Moreover, this argument plays an important role in setting up the method of investigation of the book, which is outlined in §2.4.

Third, steering people's behaviour towards becoming more sustainable without these people intending to promote the end of sustainability renders their actions instrumentally sustainable at most (see §1.2). As explained earlier, this means that, on empirical grounds, it is likely that people will return to their non-sustainable behaviour when the mechanism external to them is taken away for whatever reason. Thus, a policy that aims to influence people's behaviour without appealing to their motivation to contribute to meeting Humanity's Challenge is likely to be less effective than a policy that stimulates people's motivation for sustainable action.

The considerations of this section warrant our focus in the first instance on understanding the problem with motivation for sustainable action for individual agents and agents in institutional roles. With the above argument, we have answered the second question mentioned in §1.3: What or who should be considered the addressee of the moral requirement to take sustainable action? We proceed with discussing the methodological approach of the book in the remainder of this chapter.

## 2.3 What we can learn from psychology and its methods

We have now established the starting point for investigating the problem with motivation for sustainable action: we investigate the individual agent, who accepts Humanity's Challenge as a normative requirement for the collective of human beings, who perceives options to act sustainably within her moral horizon along with options for other morally relevant actions, and who struggles with her motivation to take sustainable action. Within this domain, we focus on the hindrances agents have to cope with. The focus is on the obstacles that agents experience within the space of moral requirements, the available options for sustainable action, and the agent's own drive to contribute to sustainability. This means that the main methodological question concerns the 'how' of investigating our key question: Why is it hard to engage in sustainable action? In other words, we must now clarify how the motivational difficulties for agents will be investigated, which is the task of the remainder of this chapter.

Different disciplines in psychology are relevant to our question. Before formulating the method of investigation, we look into the psychology of moral motivation and environmental psychology for insights that may help to describe the 'what' and 'how' of our investigation. We start with the psychology of moral motivation, followed by environmental psychology. Note that since the purpose is to find material in these branches of psychology that will be of relevance to settling on our method of investigation, the nature of this section is descriptive rather than argumentative.

The study of moral motivation by empirical psychology can be situated within the broad field of the research of human motivation. Following its Latin root (*movere* meaning movement), motivation is understood as a cluster concept that encompasses social, physiological, emotional, cognitive, and other forms of "processes that give behaviour its energy and direction" (Booth 2009, 57). The phrase "processes" emphasises that researchers disagree about what motivations are. Motivations may be understood as "psychologically compelling demands of the self, both conscious and non-conscious processes emerging from both individual and social sources" (ibid.). Thus, despite the long history of the concept of motivation, a unified empirical scientific theory of what motivation entails still seems far away (Booth 2009, 58).

Against this background, the branch of psychology interested in moral motivation investigates what prevents people from acting consistently according to their moral judgement (Heinrichs *et al.* 2013, 1). For our investigation this part of psychology is relevant, since we are interested in being empirically informed about what hampers agents from being motivated to act on Humanity's Challenge as such. What can the psychology of moral motivation tell us about the strength of the perhaps rather weak moral motive to act on Humanity's Challenge? To this end, we will, first, consider two examples of psychological theories with which the moral motivation to act sustainably can be understood. Second, we will consider a study that shows some of the motivational difficulties of acting morally. Third, to make the transition to environmental psychology, we will examine a study that approaches acting on Humanity's Challenge from the point of view of moral motivation.

For a start, like Carol Booth (2012, 413), the psychologist Fritz Oser concludes that the issue of moral motivation benefits from very few clearly defined conceptions (Oser 2013, 7). In the course of its history, the psychology of moral motivation has developed various theories about what moves people to be just, caring, and truthful. These theories formulate characteristics of moral motivation, which is understood as the force that moves an agent to act after she has morally deliberated about her action (Oser 2013, 8). Moral motivation is understood to have three features: agents act morally instead of immorally; agents consider the moral point of view, even if they do not act on it; and agents consider themselves responsible (Oser 2013, 9). Thus, in Oser's words:

> In critical situations, political, military and school leaders, but also CEOs of banks and business institutions, might or might not take responsibility for a believed necessary action that is felt under their charge. They are or are not morally motivated. If motivated, they feel accountable; if not, they feel irresponsible.
>
> (Oser 2013, 10)

Oser summarises twelve theories of moral motivation, of which we will discuss two that are suited to understanding moral motivation in relation to meeting Humanity's Challenge. Instead of developing a philosophical argument, my goal is to show that living up to Humanity's Challenge can be interpreted in terms of these theories. Understanding the moral motive to live up to Humanity's Challenge against the background of these theories provides a more detailed, psychological understanding of what motivation for sustainable action entails. Subsequently, this informs the philosophical method of investigating the agent's first-person point of view while taking a morally complicated action, which I will propose in §2.4.

The first theory puts emphasis on moral motivation as building up "a vision for the better functioning of a system and then interpret pathways and steps towards the fulfilment of that vision as absolutely necessary." According to this theory, moral motivation arises from the tension between an envisioned state of the world, which ought morally to be achieved, and the daily struggle for reaching, at least partly, this state (Oser 2013, 12). The theory argues that the agent is motivated by a feeling of insufficiency about her current situation compared with a vision of the world that is morally preferable. In the case of acting on Humanity's Challenge, moral motivation is meant to result from the tension between the envisioned moral state of a world in which future generations are able to live their lives as human beings, and where we stand today as regards taking sustainable action to further this goal, for example by being confronted with still increasing global greenhouse gas emissions (see §1.5). In the formulation of Humanity's Challenge, the moral goal of a sustainable planet for people who will live in the distant future is captured by the phrase "keeping the average temperature rise well below 2°C above pre-industrial levels". Although the envisaged moral state of Humanity's Challenge might seem abstract to many people, it can certainly serve as a starting point for a psychological understanding of the motivation of people for taking sustainable action.

Whereas this first theory emphasises the motivational force of the envisaged moral goal, the second theory focuses on the agent's "motivational ability" as a skill set required by agents if they are to persevere, to decide, to act and to be self-efficacious. Motivation, so understood, means an adaptive conscious form of searching for consistency in the situation in which the agent has to decide and act according to her moral convictions. This self-oriented form of moral motivation acknowledges that the route from judgement to intention, and from intention to action, can be hesitant, filled with delays, starts-and-stops, and many different obstacles. Motivation through self-monitoring thus means consciously ascertaining which agential capabilities are necessary for oneself in order to work towards one's moral goal (Oser 2013, 15). If meeting Humanity's Challenge is the goal to which the agent wants to contribute, then this theory holds that the agent is motivated by ascertaining her capabilities in relation to the goal of meeting the challenge. The agent has integrated the moral goal of a sustainable world for future generations as part of her practical identity. In action situations she decides what to do on the basis of her assessment of the action's contribution to her moral goal and other, contingent factors that matter to her in that situation. The agent's motivation for taking up Humanity's Challenge is understood as her ability to manage herself in overcoming any obstacles when making judgements, forming intentions, and executing a sustainable action that brings the goal a step nearer.

There appear, therefore, to be psychological theories available that can serve as a basis for explaining an agent's motivation for taking sustainable action. The (first) theory of the motivating force of an envisioned moral state is important, since it shows that the perhaps rather weak motive for taking up Humanity's Challenge (see §1.4) can be explained and detailed by means of this empirical psychological theory. The (second) theory, which focuses on the agential skill set for moral motivation, will be important as an empirical theory that underpins the motivational force of having the required agential capabilities for taking sustainable action. In Chapter 7 I will elaborate agential capabilities that are required for meeting Humanity's Challenge (§7.2).

It is clear, however, that applying these psychological theories of moral motivation does not explain why people do not act morally in many instances. I have argued why there is a need to develop a *philosophy* of the motivation for sustainable action: the interplay and accumulation ("perfect storm") of four philosophical problems hampers our motivation to act in accordance with Humanity's Challenge (see §1.3). We will now look into what *psychology* says about the phenomenon of acting on moral principles, with the goal of integrating this into formulating the method of our investigation in §2.4.[7]

In a concise paper, Batson and Thompson focus on the question why moral people do not act morally. They describe an experiment that shows how people who have well-internalised moral principles, and who are in relatively low-pressure situations, fail to act morally. They ask participants to assign tasks to themselves and to fictitious other persons. One task is clearly more desirable; it has positive payoff for the individual (the chance to earn raffle tickets). The other task has neutral consequences (no chance to earn raffle tickets) and is described as rather dull and boring. Participants are told that the other participant will not know that they were allowed to assign the tasks. The other person will think the assignment was made by chance. It turns out that most research participants assign themselves the positive-consequences task (70 per cent to 80 per cent, depending on the specific study), "even though in retrospect very few (less than 10%) say that this was the moral thing to do. Their actions fail to fit their moral principles" (Batson and Thompson 2001, 55).

By further enhancing their experiment, Batson and Thompson find two groups of participants. Some individuals want to appear moral while, if possible, avoiding the cost of being moral. The motive of these people may be called *moral hypocrisy*. On the other hand, there are participants who start with the initial motive to be moral. But when they discover that their intent to be moral will cost them the positive-consequences task, conflict arises. Then self-interest starts to overpower moral integrity. In other words, as the experiment progresses, these people's motivation may be defeated by stronger self-interested motives when being moral entails personal costs. The key difference between these groups is the agent's intent when faced with the moral situation of assigning the tasks. The former group intends to *appear* moral, yet avoiding the costs of being moral; in the latter group, the initial motive is to *be* moral (ibid.).

As Batson and Thompson suggest, from their experiment we can learn two motivational explanations for the failure of moral people to act morally: moral hypocrisy and overpowered integrity. More importantly, their research indicates that the problem of not acting morally "is not simply one of inconsistency between attitude and behaviour – between saying and doing – produced by failure to think about relevant behavioural standards" (Batson and Thompson 2001, 56). Instead, this failure has to be understood in motivational terms.[8]

In our attempt to understand the strength of the motivation to act on Humanity's Challenge, we learn from the psychology of moral motivation that two situations may occur. First, agents may want to appear to act in line with what the challenge commands; however, their motive in fact rests on moral hypocrisy. Second, although agents may want to act sustainably, they may deviate from their intention when confronted with the consequences of their action in terms of diminished self-interest (overpowered integrity).

A study that applies the methods of psychology of moral motivation to environmental psychology is that of Christian Klöckner. First, Klöckner admits that he has developed a complicated theory, which integrates various psychological theories of how people abide by their moral judgements when acting. This theory makes it obvious how far the distance between moral convictions and environmental behaviour is, and how many variables can interfere with a successful performance of morally motivated behaviour. It integrates psychological theories such as the value-belief-norm theory, the norm-activation theory, and the theory of planned behaviour, the elaboration of which falls outside the scope of this book (Klöckner 2013, 463). Second, on the basis of his model, Klöckner proposes four intervention strategies, of which two are worthwhile mentioning for our purposes. As with the theories of moral motivation, discussed above, these strategies will come up again in Chapter 7, where we discuss institutional policies that facilitate agents in taking sustainable action (§7.3).

The first strategy makes use of social role models to change people's environmental behaviour. These could be people who are significant to the target audience and who are living nearby, such as parents, neighbours, or friends. If your neighbour approaches you with the idea of recycling paper, you will be more inclined to follow this proposal than if you read an advertisement in the local news. Also, however, people who are admired by the people whose behaviour needs to change can exercise influence by showing their sustainable behaviour. In commercial and non-commercial advertising this strategy has already been in use for some time (Klöckner 2013, 466).

This strategy relies on the internalisation of social norms by the target group. "Social norms are a mental representation of the perceived or anticipated expectations of relevant people with respect to the behaviour in question – in other words, they are the internal representation of social pressure" (Klöckner 2013, 454). Social norms cause people to evaluate what significant others might expect them to do in a given situation, presupposing, of course, that these people are important enough to have their expectations considered, and that the agent is willing to give in to such expectations. This strategy will be of particular importance when we discuss the problems agents may face with their self-efficacy (§6.4). The use of role models is also important for the effectiveness of informational institutional strategies (§7.3).

The second strategy asks people to commit to environmentally friendly behaviour. The agent is asked to commit herself to perform certain behaviours in a certain time period. Expressing this commitment can either be done in private or public, orally or written. This strategy relies on the activation of already existing personal norms by strengthening the link between such personal norms and people's behaviour. Personal norms reflect what a person feels morally obliged to do in an action situation on the basis of her set of values. Personal norm theory assumes that an agent's anticipated negative effect (e.g. a feeling of guilt) resulting from a mismatch between her values and moral judgements and her actions leads to a change of action (Klöckner 2013, 456).

The strategy of commitment has shown smaller effects in the short term than other intervention techniques, but the effects last much longer. Commitment has stronger effects when it is given in written, as opposed to oral form, and when it is given in public as opposed to in private. However, this strategy has the disadvantage that people with opposed attitudes or values tend to resist committing (Klöckner 2013, 467). Throughout this book, the agent's commitment to take sustainable action will be an important topic, for which this strategy is relevant. In particular, an agent's commitment may function as a source of self-control (§6.4); also, the agent's commitment reinforces her determination to execute; for instance, when asked to sign a pledge to change their behaviour, agents are motivated to follow through (§7.3).

Whereas the psychology of moral motivation focuses on investigating the *agent*, environmental psychology studies the *relationship* between the agent and her environment. We will now consider the latter discipline.

Against the background of the growing concern for a sustainable future, environmental psychology has emerged as the discipline that undertakes "to find ways to change people's behaviour to reverse environmental problems, while at the same time preserving human well-being and quality of life" (Steg *et al.* 2013, 4). This branch of psychology investigates the factors that influence people's environmental behaviour, such as motivating factors (e.g. the weighing of costs and benefits, moral and normative concerns, emotions related to the environment), contextual factors (e.g. the availability of recycling facilities, the quality of public transport), and habits that cause behaviour (e.g. misperceptions, selective attention). An important contribution of environmental psychology is to propose interventions for encouraging pro-environmental behaviour, once the causal factors have been identified (Steg and Vlek 2009). In that sense, environmental psychology understands itself as contributing to sustainability science (Gifford 2007), to which, in the words of Robert Gifford, "psychology not only should, but must contribute more than it has so far" (Gifford 2008).

We will consider the *methods* used by environmental psychology with the aim to position the method of this book in §2.4. Where appropriate in the subsequent chapters, the *findings* of environmental psychology will be discussed in relation to the findings of this book. In this way, the investigation we undertake is carried out in relation to the fields of psychology that matter most to the topic of the motivation of individual agents for sustainable action, namely psychology of moral motivation and environmental psychology. Our goal is that this investigation of the motivational problems of agents will become empirically informed from the angle of psychology.

Environmental psychology uses five main research methods, which are both quantitative and qualitative in nature.[9] Most frequently, self-reports of individuals in response to questionnaires are used (Steg and Vlek 2009, 310). The advantage of this method is its potential for generalisation of the research outcomes to a large population, which contributes to its cost-effectiveness. Studies based on questionnaires aim to describe behaviours and to gather people's perceptions, opinions, attitudes, and beliefs about different issues. Self-reports inform us about people's convictions, as they want them to be presented; their implicit biases and their factual behaviour remain out of sight. However, the causes of those observations may be difficult to determine. For example, when questionnaires are used to find a relationship between two variables, the possibility cannot be excluded that a third variable in fact explains the relationship. Despite its statistical value, this research method may thus overlook the complexity of an action situation, considered from the agent's point of view (Steg *et al.* 2013, 7).

In contrast to questionnaires, case studies refer to an in-depth study of a particular situation. A broad topic such as motivation for mitigating climate change can be studied by narrowing down to a few agents engaged in a situation, setting or event. This research method is exploratory and qualitative in nature, meaning that possibilities for generalisation are limited. Agents are studied in their own habitat, which allows for investigating the meaning that individuals or groups ascribe to phenomena, without losing their ever-changing social, physical, historical, and cultural context. Case studies are used to develop hypotheses as input for further empirical studies. For the purposes of outlining the method of our investigation, however, it is important to note that, although qualitative research methods such as case studies are gaining in importance, "in academic journals, quantitative research methods (those that use numbers rather than words as data) still dominate environmental psychology" (Steg *et al.* 2013, 10).

In considering the use of methods in environmental psychology, two examples are worthy of note before we formulate our method of investigation. One example concerns a quantitative study; the other example is a qualitative, case study based investigation. In fact, the findings of these studies are also relevant to this investigation, and hence are included in the discussion.

In the first example, environmental psychology investigates the attitudes of people towards the environment, by asking questions like these:

> We are approaching the limit of the number of people the earth can support; the so-called ecological crisis facing humankind has been greatly exaggerated; if things continue in their present course, we will soon

> experience a major ecological catastrophe; humans were meant to rule over the rest of nature. Please indicate, by circling the appropriate response, how much you agree or disagree: strongly agree, mildly agree, mildly disagree, strongly disagree.
>
> (Koger and Winter 2010, 31)

By having sets of similar questions answered by large enough populations, statistical conclusions can be drawn regarding the attitude of people towards environmental issues. For example, the second and the fourth question refer to a worldview such as the Dominant Social Paradigm (DSP), whereas the first and the third question refer to the New Environmental Paradigm (NEP) (ibid.).[10]

The DSP can be characterised as follows: "1) Nature is composed of inert, physical elements, 2) which can and should be controlled by individual human beings 3) seeking private economic gain, 4) whose work results in progress (mostly economic development)" (Koger and Winter 2010, 38). The four assumptions of this dominant worldview in Western societies each have their own history, of which the following is an extremely brief summary (Koger and Winter 2010, 39–57). The first assumption states that in most pre-industrial cultures, nature was seen as a living, organic, spiritual entity to which human beings have a close relationship. In the course of modern history, however, this view changed to seeing nature as an orderly, mechanical, and clockwork machine. In line with this development, the second assumption invites human beings to take control over nature and to use it as instrumental to their own ends. As a side remark, what is left out of sight here is that control presupposes governance, which relates to institutions that must be in place to exercise control. The third assumption says that nothing is more important in human affairs than material reward. It forms the legitimation for capitalism to flourish. The fourth assumption emphasises the Western idea of linear progress. It is assumed that during the passage of time growth will occur, which is seen as the main metric of success. A slowdown of economic growth is seen as failure.

The DSP is in contrast to the New Environmental Paradigm, which is the view about

> the inevitability of "limits to growth", the necessity of achieving a "steady-state" economy, the importance of preserving the "balance of nature", and the need to reject the anthropocentric notion that nature exists solely for human use. Taken together, such ideas comprise a world view – perhaps best captured by the "spaceship earth" metaphor.
>
> (Dunlap and Van Liere 1978, 10)

The DSP emerged in conjunction with the rapid industrialisation of the Western world. The NEP has come about since the seventies in response to the emerging need for sustainable development (see §1.2). People tend to take their worldview as a given and accept its assumptions as obvious and commonsensical. Susan Koger and Deborah Winter therefore argue that the DSP provides a set of beliefs that encourages people to continue with their non-sustainable behaviour (Koger and Winter 2010, 36). In sum, environmental psychology tells us that dominant worldviews such as the DSP affect our motivation to act sustainably, and thus also our behaviour.

From a methodological point of view, note that the value of the method of using questionnaires is that a population's adherence to the DSP and the NEP can be measured on a regular basis. For example, it can be observed that, between 2002 and 2007, environmental concern in the US overall grew from 23 per cent to 37 per cent (i.e. environmental problems were named as the top global threat by these percentages of the research population) (Dunlap and McCright 2008, cited in Koger and Winter 2010, 33).

The second example, that of a case study based method, concerns a study of so-called lifestyle pioneers. These are people who have changed their behaviour along the lines of what sustainability demands. Lars Degenhardt looks for what motivates these pioneers, and the events in the biography of these people that have been central for developing a sustainable lifestyle (Degenhardt 2002, 124). A sustainable lifestyle is expressed by concrete behaviour – that is, specific actions oriented towards knowledge, values, and attitudes that are beneficial to the goal of sustainability (ibid., 127). Sustainable lifestyles are characterised by a moral value, which is integrated into everyday life, with the motive of personal responsibility for the state of nature and for motivating others to adhere to sustainable lifestyles. The method used in the study to investigate the motivation for choosing a sustainable lifestyle was as follows.

First, a series of interviews with a sample of people in a highly industrialised country was chosen as a qualitative method. Altogether, twelve men and ten women between the ages of 30 and 62 years were interviewed. Second, the participants were selected on the basis of four criteria. Two of these were that the person had to show a continuous and stable engagement in sustainability and that there had to be a willingness to act consistently in accordance with one's ideals or principles. Third, both a narrative and a formal interview were held with the participants. (A narrative interview has a somewhat loose format in that the interviewed person can bring up unexpected themes.) After the narrative interview, additional questions from an interview guide were asked to ensure that all the questions relevant to the study's objective were asked. The average duration of the interviews was 120 minutes. Fourth, all 22 interviews were completely transcribed, meaning that no linguistic changes or additions were made. Degenhardt reports that the willingness of all participants to report and answer questions was high (Degenhardt 2002, 128–134).

This study found that all participants were driven by consternation as their main motivational driver. They were emotionally affected by the destruction of nature and prevailing poverty; they did not want to make things worse by their actions. The actual perception of environmental damage or dangers within or outside one's own life was especially meaningful for the participants. A second characteristic of the interviewed persons was a sense of responsibility. For these people it was important to be authentic in their everyday behaviour. This meant acting according to their convictions and values. The third characteristic was a belief that taking responsibility was connected to the motive of finding a meaning in life. These people understood themselves as part of the world and took responsibility for their actions (Degenhardt 2002, 135–138).

## 2.4 The method of investigating motivation for sustainable action

As I stated at the beginning of §2.3, our investigation concerns the individual agent, who accepts Humanity's Challenge as a normative requirement for the collective of human beings, who perceives options to act sustainably within her moral horizon along with options for other morally relevant actions, and who struggles with her motivation to take sustainable action. Now that we are informed about theories from the psychology of moral motivation and environmental psychology, the set-up of our method of investigating the agent's problems with motivation for taking sustainable action will proceed in three steps. First, I explain my focus on the first-person perspective of human agency. I then outline how I approach the agent's experience in a phenomenological sense. Finally, I describe the claims of this book and their status in relation to empirical research.

Our method of investigation takes the agent's experience when acting morally as its point of departure. The agent's experience forms an integral object of study that includes cognitive, emotional, and volitional aspects. We do not consider these elements in separation. We are interested in the agent's struggle to succeed with an action that contributes to Humanity's Challenge, as seen from her own point of view. In other words, we consider a moral action, in particular a sustainable action, from the agent's *first*-person standpoint.

The choice of the first-person perspective is not an arbitrary one. As I have argued above, people in the affluent countries have some responsibility to take sustainable action (see §2.2). This means that agents raise the basic question: What am I to do about mitigating climate change? As this question forms the starting point for developing any moral point of view regarding climate change, it is thus warranted to consider moral actions as the agent experiences them. In the same way that I described agent S, who faced an unexpected reaction from her fellow homeowners (see §1.1), we shall consider these experiences through the agent's eyes. We shall attempt to find out what it is like for the agent to take action on Humanity's Challenge, while having a perhaps rather weak motive to do so (see §1.4), and while being confronted with contingencies in life that may affect the possibilities she perceives for carrying out her plan. Choosing the agent's first-person perspective as the object of study renders our method *phenomenological.*

The investigation brings to the fore the obstacles that agents experience within the space of moral requirements, the options for sustainable action available to the agent, and the agent's own drive to contribute to sustainability (see §2.3). These obstacles can be considered *empirically,* in the sense that psychology tells me, for example, that in the group of people of which I consider myself a member the DSP forms the dominant worldview adhered to. However, although there may be no doubt that this paradigm will be dominant among the people that are socially significant for me, I am sure that it is *not my* paradigm. This means that, when I reflect on my motivation to take sustainable action in my social context, it is a *normative* question for me whether I adopt the NEP instead of the DSP as the worldview that influences my behaviour. As the empirical studies show (see §2.3), worldviews tend to change among populations. The question whether I change my worldview in relation to taking sustainable action is a question of how I practically understand myself – that is, whether I want to see myself following up on the norm to meet Humanity's Challenge. Not only are the motivational obstacles for agents to take sustainable action, which will be investigated in this book, interesting from an empirical point of view, but there is a need to *philosophically* consider them as well.

We will reflect on real-life case studies with the help of the conceptual gear from relevant philosophical and psychological theories. This causes our analysis to be *conceptual.* The case studies are examples of paradigmatic sustainable actions, on the one hand, and of other morally significant actions, on the other hand. The latter ones are the so-called contrast cases.

Understanding the agent's first-person perspective against the background of concepts from philosophy and psychology means that the method of this book can be named a *conceptual-phenomenological analysis.* This should be understood as follows. The analysis is conceptual in the sense that it clarifies impersonal concepts by interpreting a real-life human being's agency in a case study. Different from empirical psychology, which is focused on describing and analysing the behaviour of people, phenomenology analyses the perspectives that people have on the phenomenon of their own experience. However, it does not do this empirically, in the way that psychology does. Phenomenology attempts to uncover general structures of perception and experience – that is, it analyses structures of people's perspectives on their experiences that can be generalised. The goal of phenomenology is to find structures of people's experience that are not necessarily impersonal, but that anyone can understand from their first-person standpoint. In that sense, the conceptual-phenomenological analysis of the cases serves as a preparation for future empirical research, for example in the psychology of moral motivation or in environmental psychology. With this investigation, our aim is to contribute to a philosophical understanding of the problem with motivation for sustainable action as far as the *individual* agent is concerned.

There is, however, another reason why we need a philosophical understanding of how moral motivation for moral action works as seen through the agent's own eyes. The problem of being concerned about meeting Humanity's Challenge and yet continuing with business as usual entails aspects of becoming conscious of one's unconscious motivations. In this regard, we can learn from psychology's phenomenon of defence mechanisms. We must be aware that, when holding agents in affluent countries accountable for behaviour that harms the environment, this could backfire through reactions that display how agents "protect themselves from discomfort, enabling them to believe that they are behaving quite reasonably".[11] Such defences are subliminal; they hinder a clear view on reality, they distort thinking, and they may eventually lead to greater environmental problems. However, they are also functional since they reduce distress in the short run.

Returning to the qualitative study on the motivational force of consternation (see §2.3), the relevance of these findings for our study is that people will not be able to creatively develop sustainable action plans unless and until they allow themselves to feel and express the painful emotions that environmental problems cause. Suppose that one has worked hard so as to be able to go for Sunday afternoon drives in a sport-utility vehicle (SUV) just for fun (see §1.2). It will be difficult to give up such accomplishments, unless agents are willing to deconstruct for themselves the meaning of high-status luxury items, while remaining a member of a community in which a culture of consumerism prevails (Koger and Winter 2010, 90).

When implementing *institutional* policies for improving sustainable behaviour in the affluent countries that happen to be constitutional democracies, the phenomenon of defence mechanisms is particularly relevant. This is so because those agents who approve the policies belong among the people whose behaviour is the object of the policy. People who populate the institutions that approve the policies thus need to be able to relate to the policies in the sense that they truly approve them. That is, the justifying reasons for the policies must be fit to motivate people to act for these reasons. Political justification of policies must take into account "the question of whether people *could* act for the reasons it offers, and … the question of what it would take for these reasons to become reasons upon which people *do* act" (McKinnon 2012, 130, emphasis McKinnon).

This means that for a policy to be effective, it must be based on an understanding of the obstacles that an agent experiences (that is, first-personally) when taking sustainable action. The introduction and preamble of the policy must be sufficiently therapeutical in the sense that it enhances the approver's recognition of her own struggle with motivation for sustainable action. An effective introduction and preamble should convince the approver that the policy not only solves a problem for the public at large, but also addresses the motivational obstacles she experiences herself.

Thus, the task of a philosophical understanding of the problem with motivation for sustainable action is to enhance the agent's practical reasoning about the question: What am I to do regarding Humanity's Challenge? As we have seen, *moral* motives[12] (i.e. what morality prescribes directly motivates the agent; or, normative reasons that simultaneously count as motivating reasons) are not sufficient to make agents act sustainably (see §1.3). The conceptual-phenomenological analysis of the agent's experience enables us to find an *indirect* motive for sustainable action. An indirect motive is one that leads to a value or good as a side effect that obtains for the agent herself. Dieter Birnbacher understands this kind of motive as follows:

> In an intergenerational context, indirect motivations do not aim at the production of goods or the prevention of evils befalling to future people but aim at objectives in the present or near future. They are nevertheless indirect motivations to act for the distant future in so far as they can be assumed to work for the good of people in the long term and to contribute to the realisation of the same ends as those underlying the principles of future ethics.
>
> (Birnbacher 2009, 285)

More specifically, in this book I will formulate an indirect motive in relation to taking sustainable action, from which the agent will benefit immediately. In this way, not only will future generations benefit from sustainable action, but the current generation of agents will benefit as well. Accepting and actively supporting policies for improving sustainable behaviour will be feasible if agents have not only internalised the moral motive for acting on Humanity's Challenge, but also found another, indirect motive from which they benefit directly.

I will take up the task of formulating this indirect motive in the final chapter. In doing so, I will make use of the conceptual-phenomenological analysis of the first-person point of view of the agent. From there, I develop the indirect motive by addressing the problem with motivation for sustainable action at the level of *humanity*. That is, for humanity – understood as all individual human beings living on earth – it matters how we understand ourselves as agents in the sense of meeting Humanity's Challenge.

With the analysis of the agent's perspective in hand, I will sharpen the formulation of the concept of moral corruption as introduced by Stephen Gardiner into the environmental ethics literature (see §1.5). The new formulation of moral corruption I propose is specifically formulated in order to address the prevailing lack of sustainable action. Thus, my understanding of moral corruption will be different from the concept of corruption of morality in general, as formulated by Gardiner. I argue that agents, not only in the affluent countries but also globally, have an indirect motive for resisting moral corruption. We cannot conceive of ourselves as persons of moral integrity if we engage in moral corruption in the way I define it and as it relates to taking up Humanity's Challenge. Since many agents are looking for ways to integrate the various roles they play into an overarching concept of their practical identity (see §2.2), the indirect motive they have for taking sustainable action is their own psychological well-being. On pain of their practical identity becoming divided, making an existential commitment to act sustainably forms the main pathway for humanity's mitigation of climate change.

Following these steps, from the point of view of meta-ethics our method of investigation should be seen as follows. We start with the moral motive to act on Humanity's Challenge, which is based on the norm to do justice to future generations. Performing the conceptual-phenomenological analysis of the first-person perspective then leads to a descriptive diagnosis of the problem with motivation for sustainable action for the individual agent. On the basis of this diagnosis an improved formulation of the concept of moral corruption is derived, which applies specifically to failing to take up Humanity's Challenge. This means that through obtaining further insight into our motivational problems with acting sustainably we find the indirect motive to prevent ourselves from getting involved in morally corrupt agency. Thus, at the end of our investigation we have an improved understanding of ourselves as acting human beings: the indirect motive for taking sustainable action not only makes us alter our behaviour to mitigate climate change but enhances our practical self-understanding as well. This enriched understanding of ourselves as agents, like the norm of meeting Humanity's Challenge, again has normative force on us.

At this point, I note that the use of the concept of phenomenology in philosophy needs explanation, given that this concept allows for a broad spectrum of interpretations (Kockelmans 2006, 666). We should therefore specify how the term is to be understood in the context of this book, given that agents act voluntarily and purposively (see §1.2).

Recall that phenomenology analyses the perspectives that people have on the phenomenon of their own experience; its goal is to find structures of people's experience that anyone can understand from their first-person standpoint. The problem with motivation for sustainable action entails aspects of becoming conscious of one's unconscious motivations. Hence, within the broad range of whatever agents may be conscious of (e.g. objects, events, other people, themselves), the basic phenomenon we are interested in is the agent, who experiences her acting and who is aware that her acting is aimed at reaching a (more or less clearly) defined goal (intentionality), and that she could have done otherwise by deciding not to perform this action (freedom). The agent acts and experiences what she does within her life world. This is the world as it is lived, that is, experienced and acted upon, and that, in turn, affects the experiencing agent (Graumann 2002, 98). To make phenomenology of action more concrete, Elisabeth Pacherie gives a non-exhaustive list of the many facets that one's experience of agency entails:

> awareness of a goal, awareness of an intention to act, awareness of initiation of action, awareness of movements, sense of activity, sense of mental

> effort, sense of physical effort, sense of control, experience of authorship, experience of intentionality, experience of purposiveness, experience of freedom, and experience of mental causation.
>
> (Pacherie 2008, 180)

Note that I do not commit myself to any specific conception either of the agential experience, or of the agent's awareness of this experience. Thereby I make sure that my discussion does not tread on something that is controversial within the philosophy of mind.

First, there is no doubt that we as human beings have causal powers of which we are aware. We act; we do things in the world out there; we are aware of ourselves as acting beings. As Tim Bayne puts it: "The experience of oneself as an agent is, for many of us, a robust and readily identifiable phenomenon" (Bayne 2008, 185). Second, we are not looking at actions that just consist of nodding one's head or lifting one's arm (see §1.2). We are considering more complex actions that require an amount of planning by the agent, that require execution, and that can be evaluated in terms of their contribution to achieving a goal the agent has set during the planning phase. Third, the actions under consideration are complicated in terms of their assessment from a moral point of view. This means that the agent in some cases, though not in all, could be faced with morally dilemmatic situations.[13]

The fact that we study sustainable action from the first-person perspective is compatible with an understanding of phenomenology as "the study of structures of consciousness as experienced from the first-person point of view" (Smith 2013, §1). The phenomena that will be studied are the characteristics of the motivating reasons of moral actions as the agent experiences these characteristics (ibid.). Therefore, phenomenology in this book should be understood, in a narrow sense, as a way of describing the experience of the agent when performing a purposive and voluntary action. As argued above, it is essential to describe the motivational characteristics of performing sustainable and other actions from the first-person point of view, since we are interested in taking these as a foundation for proposing institutional policies for sustainable action that agents can understand and support on the basis of their own experience.

An agent's awareness of the experience, while living through or performing it, makes it potentially a conscious experience (Smith 2013, §2). Note that I do not make a claim about when the agent becomes conscious of her experience. This may happen while acting, or it may happen later on when the outcome is different from what she had expected. My conception of phenomenology is "narrow" in that it does not inquire into the question how such inner awareness should be understood. For the purposes of this book, we will understand phenomenology as an interpretative method that focuses on descriptions of the experiences of the agent when performing a purposive and voluntary action. By giving a phenomenology of the motivational aspects of moral actions, we will begin to understand why sustainable actions, as a subset of this larger group of actions, seem to have their own particular difficulties.

At this point, it is useful to relate this phenomenological approach to issues in the philosophy of mind, along three dimensions. First, a descriptive account will be given of the contents of agential experience when it concerns sustainable actions and other actions that are morally significant. Since we are looking for a detailed picture of the motivations that play a role in these actions, the view given of the experience of the agents will be such as will provide a richer representation of agency than just being in control and acting with effort (Bayne 2008, 190).

Second, the agent is involved in complex action situations that require an amount of planning, that require execution, and that can be evaluated in terms of their contribution to achieving a goal the agent has set during the planning phase. This complexity, in my view, implies that the agent experiences herself as a cause. This means that the plans such an agent executes, and the high-level intentions that are involved, demand that the agent experiences herself as executing them and keeping them on track. Using the psychologist's distinction between willed intentions and stimulus-driven intentions, it seems that willed intentions are accompanied by such extensive agential experiences, whereas stimulus-driven intentions are not (Bayne 2008, 197).

Third, the agent's experience of the freedom of her agency is fundamental for our purposes. Consider a CEO who needs to close down a large factory in an area with already high unemployment. There is no doubt that this agent will experience, in the words of Galen Strawson, a sense of "radical, absolute, buck-stopping *up-to-me-ness*, in choice and action" that seems "indissociable from the ordinary, sane, and sober adult human sense of self" (Strawson 2004, 380 and 394, emphasis Strawson). Although there are enormous constraints under which the agent must operate, I assume here that the agent experiences that she can influence, at her own discretion, at least part of the course of events. We should note, however, that how this experience of freedom relates to the metaphysical reality of freedom is not the topic here (Bayne 2008, 196). Agents might merely understand themselves as agents only in the sense that they experience themselves as acting freely and that they are aware of this experience.

With these specifications in hand, we have delineated the concept of phenomenology to be used in the rest of the book. However, before we can proceed, we should clarify how the phenomenological approach outlined here relates to Scheffler's phenomenology of agency, as discussed in §1.4. Note that Scheffler's topic is the question of how individual responsibility can be understood in a world that undergoes transformations on a global scale. Scheffler argues that a consequentialist construal of responsibility seems better suited to deal with these global tendencies than does responsibility based on common-sense morality. In this context, Scheffler refers to what he calls the "phenomenology of agency" as supporting the restricted view on responsibility under common-sense morality. However, Scheffler is providing a descriptive account of how agents experience themselves with causal powers. Hence, our phenomenological approach to agency is meant as *a methodological account of what an agent's experience entails*, whereas Scheffler's phenomenology of agency *substantively* describes how agents characteristically happen to experience their acting within a limited moral world (Scheffler 2001).

We now conclude the development of our method of investigation with clarifying the descriptive diagnosis that will be developed in the following chapters.

As stated above, the starting point of the conceptual-phenomenological analysis will be agents in their individual and social contexts of action. By considering such action situations, we will look for characteristics of taking sustainable actions compared with other actions that are morally complicated from the agent's point of view. We consider the case study material (to be presented in Chapter 3) with an eye to finding out what moves the agent to choose a particular action in her specific situation. We look at the agent's experience of the motivational aspects of these actions against the background of her life world. Following up on a proposal by Karin Heinrichs for future research in the psychology of moral motivation, we consider *both* the agent's personal determinants as well as the situational determinants of her action situation (Heinrichs 2013, 651). In summary, we look at cases of individual agents taking a conceptual-phenomenological approach and follow an interpretative process that aims at understanding the phenomena.

We characterise this analysis of the agent's first-person standpoint as a *descriptive diagnosis*. Its aim is to serve as a pre-empirical hypothesis concerning the obstacles that agents encounter when having sustainable action in view. The diagnosis developed in this book, therefore, could provide a possible explanation for the motivational characteristics of taking sustainable action, which could be verified by subsequent empirical studies.

In psychology, a qualitative way of studying the experience of persons has recently emerged as an alternative to quantitative studies and is known as interpretative phenomenological analysis (IPA) (Smith *et al.* 2009). I will now briefly highlight a few of the characteristics of IPA and indicate where I deviate from this research method. As its name suggests, IPA relies on the philosophical tradition of phenomenology, supported by a theory of interpretation. IPA researchers consider significant experiences of agents and endeavour to interpret the agents' own accounts of those experiences. IPA thus relies on access of the agent to her experience and the need for interpreting the description of that experience (Smith *et al.* 2009, 3). Researchers understand that, since the agent's relationship to the world is necessarily interpretative, they have to focus on the meanings agents make out of their activities and of the things happening to them. The third element that is of major influence on IPA is idiography. IPA works with the particular situation of agents in the sense of detailed accounts of their lived, everyday experience. It also takes the context sensitivity of action situations as a starting point. As a consequence, IPA uses small, purposively selected samples, and may often use single case analyses (Smith *et al.* 2009, 21). In that sense, IPA can be seen as operating within, and attempting to extend, the philosophical current of phenomenology, in the context of psychology (Smith *et al.* 2009, 34).

There are two aspects where this research deviates from IPA. First, IPA studies done in psychology collect data by means of interviewing participants to the research and producing a transcript of the interview, which is then analysed and interpreted (see §2.3). In this book, however, the case material has not been produced by the agents at my request as a researcher, but has been obtained from the public domain – that is, literature and the Internet. Since this material has been written for different purposes (e.g. by journalists), I cannot rely on these sources as providing an account of the agent's *experience* in their action situation, as required by an IPA. This means that I have to reconstruct the experience of the agent on the basis of the material available to me.

Second, we have named our analysis a conceptual-phenomenological descriptive diagnosis. Therefore the analysis is unlike IPA in the sense that while it is descriptive in that we aim for a true record of the agent's experience, it is also evaluative to a certain extent. This means that we use the case material in the light of philosophical and psychological concepts. For example, we will be using a theory of intention from philosophy and a theory of self-efficacy from psychology to reflect on the case material in order to clarify the difference between sustainable and other moral actions. Practical reflection on the situation of the agents in the cases, with the help of philosophical and psychological theories, leads to a preliminary judgement about the characteristics of the agent's motivation while engaging in these moral actions. The conceptual-phenomenological analysis provides a way of seeing matters. It is about the operation of the imagination in bringing into the foreground the phenomenon that we want to understand – its full complexity as it is experienced.

As already explained, since I have not interviewed the agents, I have had to reconstruct the experience of the agent on the basis of the material available to me. Where the case description did not provide the necessary evidence to allow it to be reflected upon in the light of philosophical and psychological theories, I have assumed the agent to be capable of rational, autonomous agency, including action execution, and of reflection on her agency. At the same time, however, the agent remains a human being in the sense that she could be morally hypocritical, or vulnerable to overpowered integrity (see §2.3). And, like everybody else, her reactions in the action situations could be influenced by the subliminal defence mechanisms. In those instances, I have considered the agency situations by imagining the agent in that situation following the conventions I deemed applicable.

The above considerations, obviously, have a bearing on the status of the claims of this book as far as the conceptual-phenomenological analysis is concerned. I propose that the findings from the conceptual-phenomenological analysis play the role of reflective judgement, which means that they function as tentative hypotheses, to be taken as a starting point for further IPA studies

(Smith *et al.* 2009, 31). As I see it, the cases in Chapter 3 could each be taken as situations in which agents have a significant experience in their lives as agent in the personal or professional sphere. The reflections on the experiences of these agents, as developed in this book, could be considered as a tentative hypothesis that could be tested in an IPA study of similar agency situations. The purpose of the conceptual-phenomenological analysis in this book is thus not to carry out a methodologically sound IPA, but to provide a pre-empirical descriptive diagnosis of the scope and impact of problems with an agent's motivation for sustainable action. Such a preliminary judgement entails the implicit assignment to do further empirical research in order to reach more definitive conclusions (Makkreel 2015, 96).

## 2.5 The Rubicon model of action phases

On the basis of the method of investigation described in the previous section, sustainable actions and other actions will be considered that are significant from a moral point of view. However, in order to interpret the resulting action descriptions so as to form the material needed for a theory of the motivational difficulties with sustainable action, we must be able to compare these descriptions with each other. As indicated earlier, a suitable model for doing so can be found in the Rubicon model, a model that is commonly used in psychology.[14] The metaphor of crossing the Rubicon has been chosen because the model assumes that when an agent goes from *weighing* the alternatives for an action to *willing* one of them, something changes in her psychological state that seems hard to reverse. That is, the Rubicon metaphor seems appropriate since the model is based on a transition from a deliberative phase, in which the agent considers various options for action, to a firm sense of commitment (Heckhausen 1987, 6).

The Rubicon model of action phases divides a course of action into four consecutive, distinct segments or phases. It starts with a person's deliberation about her options for action and it ends with the evaluation of the achieved outcome of the action. The phenomena of committing to the chosen goal and executing the required actions are situated in between (Heckhausen and Heckhausen 2010, 3). Hence the model consists of the following phases: the deliberative phase, the commitment phase, the executive phase, and the evaluative phase. However, the fact that these phases are described consecutively does not exclude the possibility of overlap and feedback loops between the action phases. For example, in the deliberative phase, deliberation over wishes concerning a goal can easily be interrupted so that actions in the service of other already chosen goals may be planned, initiated, completed, or evaluated. While executing an action, agents may deliberate about other actions, or evaluate actions they have terminated. Whereas in psychology the Rubicon model has been developed so as "to explain the direction, persistence, and intensity of goal-directed behaviour" (Gollwitzer 1990, 55), in this book we will use the model heuristically in order to perform the conceptual-phenomenological analysis. The model enables us to learn about motivational characteristics of sustainable actions. However, it is not our objective to argue that each and every action situation can be framed into the model.

Before proceeding with our discussion, and in order to understand the Rubicon model, each of the action phases will be described briefly below. We will end the descriptions of each of the action phases by emphasising the psychological mind-set that is required for the agent to successfully accomplish the task of each phase. We follow Peter Gollwitzer's presentation of the model (Gollwitzer 1990, 56–67).

The *deliberative* phase is characterised by evaluating, deliberating, and choosing among options for action. The agent reviews a spectrum of actions on the basis of various desires she may have. People cannot usually act on all of their wishes but must choose among them, as some may contradict each other, others may be too difficult to implement, and life is simply too short to have all of one's wishes realised. In this action phase, these types of choices are made. From the point of view of psychology, these choices could be made on the basis of criteria such as feasibility and desirability. For example, with respect to feasibility, agents could contemplate whether they can obtain the outcome implied by a given action proposal through their own activity and whether the situational context they face is facilitating or impeding. Agents in this action phase will also need to consider whether they will find enough time to strive for the desired outcome and whether the necessary means or opportunities will be available.

The evaluation of the desirability of an outcome could be based on ethical considerations, or on pleasantness-unpleasantness calculations of potential short-term and long-term consequences of achieving the desired outcome for the agent herself. In the weighing of the consequences, obviously, positive or negative self-evaluations, positive or negative evaluations by significant others, progress towards some important life goal, and pleasant or unpleasant side effects may play a role. In addition, certain incentives associated with the process of achieving the desired outcome could also be relevant when the desirability of a given wish is contemplated.

In order to get through the deliberative phase, the agent should have a general open-mindedness towards processing incoming or stored information that seems relevant to answering questions about the desirability and feasibility of the options for goals to be achieved. There should thus be a cognitive attunement towards such information. Finally, the agent should also have an orientation towards accurate and impartial processing of such information.

The *commitment* phase starts when an action proposal has been adopted in the deliberative phase; the model of action phases assumes that executing the action further demands transforming the proposal to act into an intention. Phenomenologically, this transformation is characterised as a resolution that is accompanied by a feeling of determination to take the action. The desired outcome specified by the action proposal thus becomes an end state that the agent feels committed to achieving. The model describes this sense of obligation in stating that the agent has acquired a goal intention with the structure "'I intend to reach x!' whereby the x can be a behaviour or an outcome" (Gollwitzer 1999, 494).

After forming a goal intention, people move to the pre-actional stage. A phenomenon associated with this stage is planning. Since agents operate in complex environments, planning is often necessary, because newly formed goal intentions cannot be implemented immediately if the agent is engaged in alternative activities that first need to be completed or if relevant opportunities to act are not yet available. In addition, many goal intentions specify goal states that cannot be achieved in a single step (e.g. obtaining a PhD). Consequently, the agent is interrupted repeatedly and is forced to await future opportunities to work towards her goal. Note that the model of action phases assumes that agents do not use these time breaks or pauses to return to the previous phase and start weighing the positive or negative consequences of goal achievement again.

Another phenomenon in the pre-actional stage is the forming of implementation intentions. These have the structure of "When situation p arises, I will perform q" (ibid.). In this way, anticipated opportunities are linked with goal-directed responses. Implementation intentions serve the purpose of promoting the attainment of the goal specified in the goal intention. Thus, as mentioned above, the feeling of resolve associated with the goal intention makes the agent concerned with the issue of how to promote achieving the chosen goal. Accordingly, she should address questions of when and where to start acting, how to act, and how long to act. This is done by carrying out the activity of planning and the forming of appropriate implementation intentions.

The tasks to be solved by the committed agent are deciding and resolving. Thereafter, the main activities of the pre-actional stage are planning and the forming of implementation intentions in order to promote action initiation. This requires a cognitive tuning towards information that is relevant to these aspects. Second, there should be closed-mindedness in the sense of concentrating on information that helps to promote the chosen goal. Finally, following her commitment to realise the goal, the agent is expected to be moderately optimistic regarding information related to the desirability and feasibility of the chosen goal. However, this does not exclude action situations in which agents may have insufficient information, or in which they are pessimistic about whether they will be able to achieve the goal they intend.

In the *executive* phase, it matters whether a goal intention leads to initiating the relevant actions. According to the Rubicon model, this primarily depends on the goal intention's volitional strength – that is, how strongly an agent is committed to implementing the chosen goal. From the deliberative phase it follows that the actual amount of volitional strength is considered to be a positive function of the goal's desirability and feasibility as perceived prior to choosing this goal. The commitment phase enhances the goal intention's volitional strength through realistic planning and appropriate implementation intentions. Goal intentions and their effects on the initiation of relevant actions cannot be discussed without considering that many different implementation intentions may compete for execution at a given point in time. This could mean that goal intentions with the comparatively highest volitional strength are not executed automatically. The situation at hand can be such that the agent only sees opportunities to undertake actions with lower volitional strength, or the agent may ignore good opportunities to follow up on relevant implementation intentions.

There are thus four conditions that have to be fulfilled in order for a goal intention to lead to the initiation of actions that aim to achieve the goal: first, the volitional strength of the goal intention has to be large enough as compared with that of other competing goal intentions; second, the situation has to be favourable such that the particular goal intention can be readily initiated as compared with initiating competing goal intentions; third, the situation has to be favourable such that the particular goal intention can be readily initiated as compared with relevant future opportunities one hopes to encounter; and, fourth, the goal intention should have an additional advantage over competing goal intentions, assuming that the opportunity specified by the initiation intention is present. The opportunity to act should be more easily recognised and, once recognised, should elicit a special impulse to start acting on it. Action initiation is, then, the demarcation line signalling the transition to the executive phase, which is characterised by actually acting towards goal achievement. An agent's efforts to pursue a goal intention are again assumed to be related to the goal intention's volitional strength.

The task of the executive phase could be described as acting towards the goal so that goal achievement is promoted. This means that the accompanying mind-set should be one of closed-mindedness to information that could trigger a re-evaluation of the goal that is pursued, a re-evaluation of the chosen route towards goal attainment, or any self-evaluation ("Am I capable of performing this action?"). Rather, the executive mind-set should show a cognitive tuning towards internal and external signals that guide the course of action towards goal attainment.

The phenomenon associated with the *evaluative* phase involves evaluating whether striving for one's goal has succeeded. What criteria govern this evaluation? The agent must answer two successive evaluative questions. First is the question whether the intended outcome has been achieved, so that the individual may stop acting and enjoy or await the desired consequences. When the action concerns a discrete event (e.g. to send a birthday gift to a friend), this question may be easily answered. However, when the outcome can be continuously improved or extended, determining whether the intended outcome has been achieved may be difficult and full of uncertainty. Second, the agent must address the question of whether the actual value of the achieved goal matches its expected value. This implies that the agent must wait for the desired consequences of the achieved outcome before this question can be answered.

In practical circumstances, the actual value of the goal may not measure up to the expected value as assessed during the deliberative phase. The desirability of the goal may have been overestimated because certain negative consequences were neglected or underestimated, and/or positive consequences overestimated. Future predecisional deliberation should benefit from such evaluations – that is, the estimation of expected values should become more accurate. In this sense, evaluative agents not only look back into the past, but also look into the future. Moreover, post-actional evaluation may not only benefit future deliberation, but may also help an agent's future planning. It is thus the pivotal point in a feedback loop. Whenever the agent recognises that the achieved outcome does not meet the intended standards, or that the achieved outcome is not good enough to lead to the desired consequences, the agent may furnish the goal intention with new initiation and execution intentions, thus improving the chances of successful implementation.

Instead of focusing on an agent's *mental and volitional* efforts to transform a wish into the related action, the Rubicon model attempts to describe distinct phenomena of goal-oriented *behaviour*: deliberating, commitment, acting, and evaluating. In relation to this observation, three remarks are in order.

First, the model does not imply that every single initiation of action is directly preceded by deliberation about the desirability and feasibility of the underlying goal and the forming of a goal intention. Many initiations of action, for example routine household actions, are simply resumptions of activities that were started some time before; forming the underlying goal intention anew is therefore unnecessary.

Second, the model does not imply that forming a goal intention is necessarily followed by intense planning concerning where, when, how, and for how long to implement the chosen goal. It is rather assumed that such concerns originate only when smooth implementation of the goal intention is not immediately available. Action initiation may be difficult – when, for example, special circumstances are required that first need to be created, or when competing goal intentions continue to block the implementation of the goal intention at issue. Action execution may be hampered by a lack of competence on the part of the agent, or the agent's failure to focus attention on the goal pursuit when conscious control is needed. When it is unclear which condition indicates that the intended outcome has been reached, it may be difficult to decide when to stop acting.

Third, the model uses the metaphor of crossing the Rubicon to describe the forming of a goal intention. This, however, does not imply that an all-out irreversible step has been taken once the agent has formed the intention to achieve a goal. Instead, the model assumes that making a goal decision stops the agent's ongoing inner discussion. Without having new reasons to reflectively reconsider the goal to be achieved, and after the decision has been made but prior to the initiation of actions, it is assumed that no deliberation over the reasons pro and contra relative to the chosen goal is expected to occur; rather the agent is assumed to explore efficient ways of implementing the chosen goal. The model assumes that making a goal decision creates a rather durable commitment to pursue this goal, so that obstacles to one's goal pursuit do not lead to immediate retreat. Rather, the agent is expected to attempt to take hindrances away by spontaneously increasing effort, employing different means, taking more time to overcome the obstacles, or trying to get around them by taking alternative routes to goal achievement.

## Notes

1 See Hourdequin (2010) for an argument that Sinnott-Armstrong's list of principles is not exhaustive.
2 See Di Paola (2014) for another argument against the insignificance of the wasteful drive.
3 A similar comment applies to the paper by Kerri Woods, who, while arguing in favour of climate justice as "an *irreducibly collective* endeavour" (2015, 97, emphasis added), refers to "a *collective way of life* that is beyond the planet's ecological means" (2015, 101, emphasis added).
4 As Shue (1988, 696) states:

> Thus, for example, where police protection of rights to physical security is inadequate, one's positive duties will not normally be to take direct action by conducting armed patrols oneself but to pay higher taxes in order to support a larger or better police force. The direct duties involved in maintaining enough law and order to protect physical security fall upon such institutions as police forces and judicial systems, but individuals generally have various indirect positive duties to support the institutions that bear the direct duties.

5 It could be objected that not living up to the obligations assumed under the UNFCCC primarily poses a *legal* problem. Note, however, that although the Paris Agreement is considered a treaty under the Vienna Convention on the Law of Treaties, some scepticism is warranted that its status under international law will in fact influence the behaviour of states. The UNFCCC lacks enforcement mechanisms and it seems unlikely that courts will apply it at the international or at the domestic level (Bodansky 2016). Whether it is legally binding or not, in the Paris Agreement states have agreed to live up to certain obligations. As it is entirely clear that indigenous peoples and the developing countries are least responsible for climate change, least able to adapt to it, and most vulnerable to its impacts, the obligations assumed under the Paris Agreement count as moral obligations.
6 The IPCC (2014b, 15) estimates that a reduction of global anthropogenic greenhouse gas emissions so that they are between 40 per cent and 70 per cent below 2010 levels by 2050 is necessary to achieve the global warming limit of 2°C during the twenty-first century.
7 In psychology of moral motivation this question has been studied already for decades, and it is beyond the scope of this book to give even a summary of this field of research. Asch (1956) and Milgram (1974) form classic examples of important papers in this regard.
8 Monin and Merrit (2012) have broadened the scope of the research by Batson and Thompson to include the difference between hypocrisy without inconsistency and inconsistency without hypocrisy.
9 See for a more detailed description of research methods: Steg *et al.* (2013), 6–10. Note that I leave out a description of laboratory experiments, computer simulation

studies, and field studies, since these are less relevant for the positioning of our research method.

10 Measuring people's general attitude towards the environment has been done since the seventies, e.g. through use of the Ecological Paradigm Scale (Fransson and Gärling 1999, 370).

11 Koger and Winter (2010, 72–79) explain defences such as rationalisation, intellectualisation, displacement, suppression, repression, denial, reaction formation, projection, apathy, and sublimation.

12 Birnbacher (2009, 281) defines the different kinds of motive (emphasis Birnbacher):

> A morally required act is done from *moral* motives if it is done precisely because it is morally required, i.e. from conscientiousness or a feeling of duty. It is done from *quasi-moral* motives if it is done from altruistic motives such as love, compassion, solidarity, generosity or spontaneous impulses to care for others, i.e. from motives that often lead to the same courses of action as genuinely moral motives, without being dependent on the adoption of a particular system of morality. … Non-moral motivations comprise both self-centred and non-self-centred motivations that result in morally required action accidentally, such as the desire for self-respect, social integration and recognition, and the pursuit of personal ideals from which others happen to profit. (These are only ideal types. In reality, there may be all kinds of combinations of these kinds of motive.)

See also Birnbacher (2018).

13 This means that "the agent is required to do each of two (or more) actions; the agent can do each of the actions; but the agent cannot do both (or all) of the actions" (McConnell 2014, §2).

14 For a discussion of the Rubicon model versus other models of action in psychology, see Gollwitzer (1990).

# 3 The case study material

## 3.1 Introduction

We have chosen the group of affluent countries as the moral community that ought to take up Humanity's Challenge (see §1.5). Given that these countries are constitutional democracies, I have argued that individuals bear some responsibility for taking up the challenge by starting to act sustainably (see §2.2). In the remainder of Chapter 2, I have developed the method of the book, which can be summarised as performing a conceptual-phenomenological analysis in order to arrive at a descriptive diagnosis of the motivational difficulties with taking sustainable action. The analysis undertaken involves clarifying the motivational issues by understanding agency situations from the agent's first-person perspective with the help of concepts from philosophy and psychology. While the philosophical and psychological theories will be introduced within the chapters of Part II that deal with the Rubicon action phases (see §2.5), the purpose of this chapter is to provide the case material in the form of the accounts of agents who take sustainable and other morally complicated actions.

I consider that, as agents in constitutional democracies, these case study agents are, first and foremost, citizens of their societies. Following Niklas Luhmann, and given that they have evolved towards a state of affluence, we may consider these societies as functionally differentiated. This means that societies must fulfil necessary functions that are interdependent, such as the political function, which provides for collectively binding decisions, and the economic function, which secures the satisfaction of certain needs over time (Luhmann 1977, 35). In my view, while being citizens, agents perform roles both as individuals and as institutional role occupants. That is, I understand agents taking up roles as individuals if these roles are initiated from within their lives as a private person; they perform institutional roles when they, for example, exercise a role as a professional within the context of a social institution.

Thus, the case material consists of examples of agents, in both individual and institutional roles, who take sustainable actions and actions that are morally relevant in different ways. The other morally relevant cases are included so that they can be analysed in contrast with the examples of sustainable actions. This, however, does not imply that we are looking for unique characteristics of the motivational difficulties with taking sustainable action. We are investigating features of sustainable actions that could be helpful in preparing individuals for overcoming these motivational difficulties; at the same time, these features could be of use to institutions that aim to implement policies that facilitate the taking of sustainable action.

As to the format of the case material, the following remarks should be made. First, the material from which the case studies are derived is available in the public domain and can be accessed by the reader. Second, an effort has been made to obtain accounts of morally relevant actions from the agent's first-person point of view as far as possible. In a proper IPA study, this would require that the researcher interview the agents or, at the very least, that the agents leave behind autobiographical notes concerning their actions, including any retrospective comments about their actions from their own point of view. In this investigation, however, these strictly first-person accounts, could only partly be retrieved in instances where, for example, the agent was sufficiently well known to be interviewed by a journalist, or where the agent gave an account on the Internet via a blog or a dedicated website. In other cases, the case description has been constructed from the available material. There is, thus, a small element of fiction in the case materials provided. For our purposes, however, that is not objectionable; the case studies suffice for getting started on the kind of reflection we engage in in ensuing chapters.

A disclaimer about the completeness of the case study material should be made, however. The purpose of the case study material is to provide accounts of morally relevant situations of agents in affluent countries. The case descriptions do not pretend to provide full coverage of all relevant facts. Therefore, the reader who is already familiar with the details of a particular case study will see that certain features are almost certainly not reflected in the description given below. Nevertheless, given the goal of this book, the case descriptions are intended to be realistic and compatible with more detailed accounts.

In the following chapters, the conceptual-phenomenological analysis is developed using the five cases of agents in individual and in institutional roles as presented below (§§3.2–3.4 and §§3.5–3.6 resp.). The cases of agents in individual roles are more or less confined to the private sphere – that is, the sphere where agents perform a non-professional, non-institutional role. For example, individual agents could be the father or mother of a family, a board member of the local tennis club, someone who takes her friend out for dinner. In total three cases are presented, the first and third case (§3.2 and §3.4 resp.) being examples of sustainable actions that help meeting Humanity's Challenge, the second case (§3.3) being an example of an action that is morally relevant for other reasons. Subsequently, two case studies are presented of agents performing their institutional roles as employees of corporations in the economic field. In order not to introduce unnecessary complexity, we assume that these professionals do not occupy managerial roles. One case concerns a sustainable action (§3.5) and the other one is morally complex for different reasons (§3.6). In the final section of this chapter, I explain the consequences of integrating the method of investigation (see §2.4) with the Rubicon model of action phases (see §2.5) in the form of a descriptive diagnosis for the structure of the ensuing chapters (§3.7).

## 3.2 The semi-vegetarian

As an example of a sustainable action, consider someone in a high-income OECD country, who decides to become a (semi-) vegetarian. In line with Humanity's Challenge, the agent has come to the moral judgement that drastically reducing her consumption of meat from mammals contributes to reducing the global scarcity of fresh water and that it, thereby, counts as a sustainable action (Mekonnen and Hoekstra 2010). What are the obstacles the agent encounters in carrying out her decision? Here are some examples.

The agent must change her cooking habits. Meat is not on the menu anymore; hence, new ways of cooking have to be developed. Before the agent discovers what she *can* eat as a vegetarian, she realises what she *cannot* eat anymore.

Buying food becomes more complex. Products that are being sold "without meat" are not necessarily vegetarian. Food products may contain animal oils or fats. Reading labels on food products and acquiring knowledge about food production is required when the agent is determined to carry out her decision.

The action seems to rule out going out to steakhouses for dinner. The choice of restaurants becomes limited. When going out for dinner with friends or colleagues, one's decision to be a vegetarian could jeopardise the social atmosphere in the group. It could be perceived that the agent's *individual* choice to be a vegetarian imposes limits on the decisions of the others in the *group*.

And what is to be done about dinner engagements? It seems that having decided to become a vegetarian leads to a situation in which the agent's friends are restricted in their choice of dinner, which they in turn want to offer to their vegetarian friends. In other words, *my* decision to be a vegetarian results in restrictions on *you*, who voluntarily want to offer me dinner as a friend.

Do vegetarians want to convert me? Even without undertaking any attempt to convince others of (semi-)vegetarianism, simply acting according to one's normative viewpoint potentially arouses a discussion about the question whether others should convert to the same normative viewpoint.

## 3.3 The doctor without borders

Here is an example of a different type of morally complicated action. After a career of twenty years as a surgeon, Tom Krueger decides to join Médecins Sans Frontières (MSF 2012). Krueger lives with his family on a 40-acre farm in Tennessee (United States). He says:

> I was tired from what I was doing. I was fed up with the US healthcare system, whereas being a surgeon has always been my calling. Becoming a doctor without borders, however, means a lot to you and your family. My decision had many consequences. Therefore, it has taken me twenty years to join MSF.

Being on an MSF mission is hard. You experience things you cannot share with any of your friends. Under these circumstances, there is hardly any distinction between what is professional and what is private. What you can do is of more significance than you are used to. However, due to shortage of almost everything, you know that people die, for example while you are having lunch. You are constantly asking yourself whether someone else would accomplish more than you do. Being on mission means a shock to your Western values. If you cannot take decisions, you are not up to the job. So, you have to live with the fact that you occasionally take the wrong decisions.

Tom Krueger says:

> I feel good about what I am doing. I want to do it better. Fixing other people seems to fix yourself. So, that is how I fix myself. That's what I found out. Sometimes you have to leave what you are doing behind in order to find out what is missing.

He has served on MSF missions to Darfur, Nigeria, and Sri Lanka.

## 3.4 The "power grid rebels"

Following the Chernobyl disaster on 26 April 1986, a group of people in the German village of Schönau (2600 inhabitants) decided to take sustainable action under the slogan "Parents for a nuclear-free future" (Janzing 2002, 272–275). These people did not want to wait any longer for politicians, power grid suppliers, and other industrial companies to do something to address the dangers of nuclear power; they wanted to take action themselves.

In 1990 matters became serious. The 20-year electricity concession between Schönau and the regional power company KWR was due to expire at the end of 1994. In order to protect its interests, KWR offered the municipal authorities four yearly payments of 25,000 Deutschmark (DM) to secure the immediate renewal of the concession. In the eyes of KWR, a discussion about the possibilities of moving towards a more sustainable power supply for the village could be prevented in this way. However, such a discussion was exactly what the civilian initiative was hoping for. Their aim was to establish sustainable power delivery for Schönau. Given the financial attractiveness of the offer by KWR, and the questionable feasibility of the ecological alternatives at that time, the prospects for the civilian initiative were dim.

The civilians did not give up. In November 1990 they institutionalised their initiative by founding a company to buy back the power grid of the village. This company promised the municipality that it would match the offer by KWR, through fundraising among private individuals, if the negotiations about the concession could be delayed. The available time could then be used to thoroughly discuss the alternatives for the power supply by KWR. And, indeed, the initiative managed to mobilise 280 people who raised the required DM 100,000. Even when KWR increased its offer to DM 55,000 in 1992, the civilians were able to raise the required funds to match this offer. Despite this, with a majority of one vote, the municipal council decided in 1991 to reject the offer by the civilian initiative and to renew the concession with KWR.

Yet again, however, the civilians, led by Ursula and Michael Sladek, did not give up. They made use of the regional constitution that allowed for the organisation of a referendum, by which the inhabitants of Schönau could vote for or against a sustainable power supply. With a turnout of 75 per cent of the inhabitants, on 28 October 1991 the initiative won the referendum with a majority of 56 per cent of the votes. The outcome of the referendum resulted in the requested delay to the negotiations with KWR, which allowed Mrs and Mr Sladek, who in the meantime had gained countrywide publicity and been nicknamed "Stromrebellen" ("power grid rebels"), to create support for their plans for an ecologically friendly power supply for their village.

In the months that followed, it became clear that KWR would not accept the demand for an environmentally friendly power supply. This led the rebels to develop plans to take over Schönau's power grid from KWR. Obviously KWR wanted to prevent such a takeover, not because it would lose 2600 customers, but because of the precedent that it would set. The legal situation, however, was in favour of the civilian initiative: if the municipality were to refuse to renew the concession with KWR, the energy company would have to sell the power grid back to the municipality. By setting a sales price of DM 8.7 million, KWR tried to scare away the civilian initiative, which, in contrast, had obtained professional advice that the village's power grid was worth only DM 3.9 million.

The initiative decided to start legal proceedings against KWR to achieve a lower price. However, as such a procedure could easily have lasted for ten years, the rebels decided to make the down payment of DM 8.7 million, wait for the legal proceedings to conclude, and claim the surplus (i.e. the difference between DM 8.7 million and the price to be decided by the court) back from KWR later on. By doing this, the village could start its own electricity company with immediate effect. In any case, the people driving the initiative were becoming impatient with the large amounts of money they had to raise through gifts and contributions from many other people.

In January 1994, the Electricity Company of Schönau (EWS) was established. It acquired the concession on 20 November 1995. The opponents of

EWS, however, had learned that a referendum could result in a victory, and now it was their turn to use this legal instrument to reverse the decision of the municipal council. The second referendum was held on 10 March 1996. Once again, the "power grid rebels" won, with a record participation of 85 per cent and a majority of 52.4 per cent of the votes.

Ultimately, the conflict about the sales price of DM 8.7 million had to be settled. Another consultancy advised a price of only DM 6.5 million, and KWR was forced to accept it. Although EWS was of the opinion that the DM 6.5 million was still too high, it started to raise the required funds. The fundraising campaign benefited from the support of a much larger group of people than the inhabitants of Schönau only. Finally, KWR, whose reputation in society had been severely damaged, helped EWS succeed by lowering the price to DM 5.7 million. On 1 July 1997, EWS took over the power grid of Schönau from KWR. The civilian initiative had reached its goal. When the German energy market was liberalised in 1998, EWS was able to sell electricity through the entire country. Approximately 12,000 households in Germany changed their power supply to the sustainable electricity producer in the Black Forest.

## 3.5 The purchase manager

Palm oil is widely used in many consumer products, such as margarine, chips, biscuits, and soups. On a trip to Indonesia and Malaysia in 2007, the purchase manager for oil and fat of FrieslandCampina, one of the largest dairy food companies in the world, learned about the abuses on palm oil plantations (Gigengack 2013). Large parts of the Borneo jungle were being cut down illegally, inhabitants of the jungle were being expelled, and biodiversity was being destroyed by deforestation. The purchase manager concluded that things were not going the right way.

She realised, however, that virtually nobody in the headquarters of her company would support an employee putting these abusive situations on the corporate agenda. But, she felt, she could not walk away. She said: "I did not want to wait for more stringent rules; I wanted to take action myself."

The purchase manager started by looking for environmentally friendly alternatives. This did not work out. In her view, not only should the suppliers be prevented from cutting down the jungle, but the social circumstances in the entire production chain should improve as well, for example by allowing unions in Indonesia. She thus started discussions with her senior managers about the company becoming a member of the Roundtable on Sustainable Palm Oil (RSPO). The purchase manager met with clear scepticism. A typical retort was: "Why should we put time and money into this, if our clients do not ask for it?" All parties in the palm oil production chain were pointing at each other; there was always an excuse not to take sustainable action. The purchase manager felt frustrated about this deadlock.

However, she did not give up. She persisted in trying to bring the unsustainable practices in palm oil production to the attention of the senior executives in FrieslandCampina. Finally, the purchase manager was given the opportunity to present her plans to the CEO of the company. He was convinced almost immediately and supported the plan. Consequently, FrieslandCampina started requiring an RSPO certificate from its palm oil suppliers.

With hindsight, the purchase manager was able to reflect: "The most difficult obstacle to overcome was to align the parties in a complex supply chain. In order to achieve the goal of sustainable palm oil production, everybody has to cooperate, whereas not everybody has the same interest." She also stated: "By holding on to one's vision, an individual can make the difference in reaching a goal like this. However, if I had not been committed to the topic of sustainable palm oil, I would not have persisted."

## 3.6 The credit restructuring officer

As an example of a different type of morally complicated action in an institutional context, consider an employee in a bank who is in charge of credit restructuring. Usually this involves dealing with borrowers of the bank who have breached a clause in their loan agreement in such a way that the interests of the bank are in jeopardy. Assuming that the borrower is acting in good faith, there can be reasons for their not being able to act according to the contract. In many instances, the client's actual revenues have fallen below the turnover that was projected at the time the loan agreement was entered into, or the client's operations are organised inefficiently, resulting in unnecessarily high costs. In a worst-case scenario, these predicaments result in the borrower no longer being able to pay interest and repay the loan.

Given that her task is to propose and take actions that will improve the situation in the interests of the bank's stakeholders, the restructuring officer is confronted with the question as to how the interests of the parties involved should be weighed. On the one hand there is the alternative of putting the client's interests first – that is, continuing to finance the borrower's business, whilst taking a loss in the form of unpaid interest or accepting a part of the loan amount not being repaid. If this way of acting were to become common practice in the bank, this would potentially negatively affect the interests of other stakeholders, namely deposit holders and shareholders. It might even lead to the bank's bankruptcy, as has been observed during economic downturn periods, for instance in the early 1980s.

On the other hand, the restructuring officer could honour the interests of deposit holders and shareholders by ending the credit relationship, ensuring that taxpayers will bear the cost of the borrower's bankruptcy. By following a legal procedure, the bank will declare the borrower bankrupt, an executor will replace the entrepreneur, and, insofar as the borrower's assets are pledged to the bank, these will be sold in order to recover interest and the principal loan amount. If the bank were to pull the plug on some restructuring cases at a very early stage causing substantial unemployment in an industry sector, this would put the bank's reputation at risk as a trusted party that cares for its clients and guards them even through economically difficult times.

In summary, the situation of the credit restructuring officer can be considered morally dilemmatic: honouring the principle to put the client's interests first could lead to overruling the legitimate interests of other stakeholders (deposit holders and shareholders); leaning too much towards the interests of the other stakeholders endangers the bank's reputation, and – in a competitive market place – could lead to clients leaving the bank, which, in turn, negatively affects the bank's own income line. The agent, therefore, is required to perform both of these actions; she can do either of them, but she cannot do both. It is, therefore, warranted to consider the job of a credit restructuring officer in a bank a morally non-trivial one.

## 3.7 Programme of Part II

Having now completed the preliminaries of Part I by presenting the case study material in this chapter, the programme for the remainder of this book will be as follows. We move on in Part II to integrate the phenomenological approach to agency (see §2.4) with the Rubicon model of action phases (see §2.5). This means that for each of the phases of the Rubicon model the case study material is subjected to a conceptual-phenomenological analysis as outlined above. As part of this analysis, sustainable actions will be compared with actions that are morally complicated for different reasons.

Thus, starting in Chapter 4, the deliberative phase will be assessed for the cases described in Chapter 3. Following the phenomenological approach, we will perform an analysis of the cognitive issues that come about in this phase and distinguish between sustainable actions and other actions. In Chapter 5, for the commitment phase, the specific features of taking decisions in favour of morally complex actions and planning the execution of those actions will be analysed. Likewise, in Chapter 6, the features of executing sustainable actions are compared with execution of other morally complex actions. Note that relevant aspects of the evaluative phase will be integrated into Chapters 4–6 along the way.

# Part II

# A conceptual-phenomenological analysis of the Rubicon action phases

# 4 Motivational challenges in the deliberative phase

## 4.1 Introduction

In accordance with the Rubicon model, we assume that an individual agent's action starts with the deliberative phase. In this phase, the agent is engaged in practical reasoning, which entails assessing, weighing and choosing among options for action. We assume that the agent does not question Humanity's Challenge, but takes it to be the normative background assumption for human agents *collectively* to take sustainable action. Note that we have argued from a methodological point of view that the investigation should begin by considering the agent's first-person perspective (see §2.2). However, this does not mean that the agent accepts that she is under the duty as an *individual* to act on Humanity's Challenge. Given this, we attempt to answer the question: What happens in the deliberative phase that could hamper the agent's progress towards the next action phase? In other words, what are the main impediments, while she is assessing, weighing, and choosing among options for action, that could hinder the agent in moving towards the commitment phase?

As the description of the deliberative phase suggests (see §2.5), the range of reasons that could hamper the agent in choosing a particular action from among other options for action is immense. There could be other plans the agent is executing that prevent her from continuing with the one she is contemplating. Deliberating about prospective actions, for whatever reason, could arouse strong memories and accompanying emotions in the agent that block her from progressing further. Life may be simply too hectic for the agent to feel capable of finishing the process of assessing, weighing, and choosing to the point at which the agent normally takes these types of decisions. Therefore, in order to create focus in this chapter, we start with exploring the deliberative phase. We discuss the kind of deliberative questions the agent faces, and we provide methodological precision concerning the practical syllogism and the Rubicon model. We continue with discussing some examples of impediments that could play a role in this phase without having any pretension to give a complete overview (§4.2).

Having thus broadly determined our focus, we continue with discussing the two remaining philosophical obstacles of the four that were introduced in §1.3.

The third obstacle is contained in the question: From the point of view of common-sense morality, what can be expected of the agent as regards acting in conformity with the norm to meet Humanity's Challenge? This leads to a debate about the question whether ethical theories (e.g. a normative theory promoting sustainability) lead to demands on agents (e.g. to eat less meat, to install solar panels) that appear to be unreasonable from the agent's point of view, because they are too demanding in the eyes of the agent. As a result of respecting the agent's discretionary moral space, we might simply see no way to carry out the complex coordination required to achieve a global reduction of greenhouse gas emissions as demanded by Humanity's Challenge (§4.3).

The fourth philosophical obstacle we encountered in §1.3 can be derived from the question: What can be said about the motivating force of the obligation to respect the rights of future generations? A motivational problem arises if the agent argues: I delegate my decision to take sustainable action to a future self. Therefore, I don't need to work on Humanity's Challenge now. The underlying conception of the human being that the agent may be tempted to adopt here is Parfit's Reductionist View.[1] This view says: The identity of future people is fully contingent on the actions of current people. If future people are born, they will owe their very existence to the actions that I will have taken; hence, it does not matter whether I act sustainably now. In response to this challenge, I will formulate a normative-phenomenological conception of human agency that we must necessarily presuppose if we want to avoid the temptation of delegating our task to take up Humanity's Challenge to a future self. I will argue that this conception of human agency is incompatible with the Reductionist View. Since we have an alternative conception of human agency, and since the moral implications of the Reductionist View are problematic, we ought to resist temptations to adopt the Reductionist View – so will I argue (§4.4).

We end the chapter by giving an overview and diagnosis of the motivational challenges in the deliberative phase (§4.5).

## 4.2 Exploring the deliberative phase

The deliberative phase starts with the assumption that the agent has accepted Humanity's Challenge as a normative reason for human agents *collectively* to take sustainable action. This means, first, that although the agent accepts that it is up to humanity as the collective of currently living human beings to act on Humanity's Challenge, she has not yet made the challenge a personal challenge for her. In particular, she has not yet assumed individual responsibility for taking up Humanity's Challenge. Second, however, the argument for weak internalism (see §1.4) entails that the agent has a non-negligible, but a perhaps rather weak, motive to take action in order to meet the challenge. Thus, although it seems plausible to assume that the agent has *a* motive for deliberating in favour of doing the sustainable action, this motive may be easily overruled by other, stronger motives. Third, as regards the moral community that ought to take up Humanity's Challenge (that is, agents in the affluent countries, see §1.5), it can be expected that this community is capable of taking the normative reason entailed in Humanity's Challenge as a motivating reason for sustainable action – that is, a reason that explains their behaviour.[2]

Given the assumption of Humanity's Challenge as a collective action problem, and what follows from it, we now consider agent deliberation about options for sustainable action in relation to deliberation about other actions that are morally relevant for different reasons. We are looking for the motivational features of sustainable actions in the deliberative phase that hamper the agent in proceeding with her rather weak motive to take up Humanity's Challenge towards the commitment phase. From here, the exploration of the deliberative phase will proceed in two steps. First, we will discuss the problem of collective action and elaborate a more precise formulation of what deliberation entails by relating the practical syllogism that explains why an agent acts as she does to the Rubicon model of action phases. The analysis of the practical syllogism in relation to the Rubicon model reveals an important methodological precision in the demarcation between the deliberative and the commitment phase. Second, we will look into the empirical research concerning the motivational obstacles for sustainable action that can be assigned to the deliberative phase. An overview will be given of some relevant empirical findings.

Since it affects literally all human beings, the problem of meeting Humanity's Challenge can be considered among the most complex collective action problems in the history of mankind. Along with Stephen Gardiner, one could phrase this problem as a so-called tragedy of the commons:

> Roughly speaking, the tragedy of the commons holds when: (TC1) each agent prefers the outcome produced by everyone restricting their consumption over the outcome produced by no one doing so; but (TC2) each agent has the power to decide whether or not she will restrict her consumption, each (rationally) prefers not to do so, whatever the others do.
>
> (Gardiner 2011a, 27)

Suppose that climate change were caused just by driving in SUVs on sunny Sunday afternoons (see §1.2). If we view the atmosphere as a commons, TC1 tells us that we would prefer the outcome of a climate not polluted by greenhouse gases emitted during such drives. However, according to TC2, agents are tempted to consider the polluting effect of their individual drive insignificant, and thus prefer to go for their drive, whatever other people do. If agents agree with the first claim, but act en masse according to the second claim, then that situation generates the tragedy of the commons.

The agent's deliberation in this phase of the Rubicon model takes place against the background of the tragedy of the commons, as laid out by Gardiner. The agent has accepted that meeting Humanity's Challenge is the preferred outcome for humanity as a collective. Since the agent cannot reasonably deny that she is a member of the moral community that ought to take up Humanity's Challenge (see §1.5), this gives her a normative reason – a reason that justifies her action – to choose a sustainable action among the options for action she has at her disposal. As already stated, this does not mean that she accepts individual responsibility and will act accordingly. However, the agent cannot persistently refuse to take notice of the fact that she has *a* reason that justifies her deliberation in favour of taking a sustainable action.

The agent's deliberation is an instance of practical reasoning. As Paula Gottlieb phrases it: "Deliberation, then, occurs when everything is not already set in advance, and yet it is not completely arbitrary what to do" (Gottlieb 2006, 220). From accepting certain premises, the agent moves towards taking a concrete action. Her deliberation starts from the premise that she has a normative reason to take up Humanity's Challenge. That is, the justifying reason to help in meeting Humanity's Challenge forms the major premise of a practical syllogism:

| | |
|---|---|
| *Major premise:* | I have a normative reason to contribute to efforts to meet Humanity's Challenge. |

A practical syllogism explains an action by specifying the reasons for which the agent takes the action. This means that next to the normative reason as entailed in the major premise, a second premise has to be added in order to lead the agent to act. The second premise must contain a belief that through executing an action with certain characteristics the agent actually lives up to the justifying reason or attempts to do so. A practical syllogism looks like this:

| | |
|---|---|
| *Major premise:* | I have a normative reason to act in accordance with φ. |
| *Minor premise:* | I believe that through executing an action with characteristics X, φ will be realised or its realisation will be furthered. |
| *Conclusion:* | I intend to execute an action with characteristics X now.[3] |

By substituting the normative reason to help meeting Humanity's Challenge, we obtain the following practical syllogism, which forms the topic of this chapter:

**Practical Syllogism for Sustainable Action (PSSA)**

| | |
|---|---|
| *Major premise:* | I have a normative reason to contribute to efforts to meet Humanity's Challenge. |
| *Minor premise:* | I believe that through executing an action with characteristics X, meeting Humanity's Challenge will be realised or its realisation will be furthered. |
| *Conclusion:* | I intend to execute the action with characteristics X now. |

In other words, the agent's deliberation in this phase is about the minor premise – that is, about the evaluation of options for action that contribute to meeting Humanity's Challenge. We are investigating the obstacles to identifying the minor premise that enables the agent to move towards the conclusion. Therefore, investigating what could hamper the agent from proceeding from the deliberative towards the next action phase means that we investigate which competing motives the agent could have that are capable of overriding the motive to deliberate in favour of the sustainable action.

However, before we continue with this investigation in the remainder of this chapter an obvious question is: How does the practical syllogism relate to the Rubicon model of action phases, which forms the methodological framework of the book? Notice that the syllogism should be considered in relation to the many other events occurring in the agent's daily life. A range of other motives that the agent wants to follow could affect the formation of a syllogism that leads to a sustainable action. Her identification of a suitable minor premise may be disturbed by another motive she wants to act on, such as planning for an extra holiday, which may conflict with the motive of placing solar panels on the roof of her house (e.g. see §1.1). Moreover, the agent's reasoning about the appropriate minor premise becomes truly practical as the agent considers the practical skills that are involved – that is, the agent's practical skills in terms of assessing the characteristics of the action in as far as it contributes to meeting Humanity's Challenge and the skills needed to execute the action form part of her deliberation about which action to take (Gottlieb 2006, 218).[4]

At this point, we can specify further the transition from the deliberative phase to the commitment phase by looking into Davidson's account of practical reasoning, which can be expressed as proceeding from prima facie judgements to all-out judgements (Rödl 2007, 25–27). Prima facie judgements express a desire for taking the action. That is, they entail an attribute that counts in favour of doing the action (Davidson 2001, 87). However, such judgement does not yet mean that the agent has concluded that the action must be executed all things considered. This part of the practical reasoning, i.e. proceeding from a number of prima facie judgements to an all-things-considered judgement, belongs to the deliberative phase. I thus take it that the deliberative phase ends with the all-things-considered judgement that refers to "things known, believed, or held by the agent, the sum of his relevant principles, opinions, attitudes, and desires" (Davidson 2001, 41). With the all-things-considered judgement the agent expresses that she accepts that from a number of possible actions X is the one to be executed.

Thereafter, the commitment phase starts with the agent making her commitment in the form of the all-out judgement. The all-out judgement is an unconditional judgement about what the agent wants to do; it is the intention to execute the action (Davidson 2001, 88). We can thus conceptualise the demarcation between moving from the deliberative phase to the commitment phase as lying in between the all-things-considered judgement and the all-out judgement.[5]

In Chapter 5, we will see which motivational obstacles agents encounter when they proceed through the commitment phase. Executing the action, which follows from the intention as concluded in the practical syllogism, forms the executive phase of the Rubicon model, considered in Chapter 6. By arguing from Davidson's account of practical reasoning, thereby specifying the point of change from the deliberative phase to the commitment phase, we have thus aligned the Rubicon model with the practical syllogism. At first sight, the commitment phase seems somewhat hidden in the practical syllogism, but it comes to the fore when we focus on the agent while she practically deliberates and proceeds from the all-things-considered judgement to the all-out judgement.

We will now consider examples of motives that compete with the one we have argued for: the agent having an interest to help meeting Humanity's Challenge. In following this motive, we consider the agent, whose task it is to choose from a range of possible acts the one that, all things considered, is the most sustainable. Since she has chosen the end of her action to be meeting Humanity's Challenge, doing this act will render her action sustainable.

From the psychology of moral motivation we have learned that there are potentially many instances in which people do not act on their moral principles (see §2.3). By considering the practical syllogism above, we observe how the deliberation about options for sustainable action (minor premise) can be affected by competing motives. The motivating reason for taking the sustainable action (the reason that explains the agent's behaviour), then, is potentially overridden by other motivating reasons. This means that the way the agent evaluates the means she takes to realise her ends influences her efficaciousness. The agent's habits of evaluating options for action could lead her to conclude that acting morally on a particular occasion might not be a suitable means to her ends (Kolodny and Brunero 2013, §1).[6] That is, the agent's habits of evaluating options for action may prevent her from acting sustainably. However, since the agent's deliberation is practical, she is responsible for her reasoning, in particular for her habits of evaluating options for action.

The agent proceeds from a number of prima facie judgements to an all-things-considered judgement, on the basis of things known, believed, or held by her, the sum of her relevant principles, opinions, attitudes, and desires. In the next paragraphs we will consider situations in which the agent's practical deliberation does not lead her to act upon the motive to help meeting Humanity's Challenge. The agent does not effectively pursue her aim as accepted in the major premise of the practical syllogism. This happens when she intends to contribute to meeting Humanity's Challenge but fails to be instrumentally coherent – that is, she believes, roughly, that something is a means to act upon the motive, but refuses to correspondingly intend, or is hampered in thus intending (Kolodny and Brunero 2013, 10).

To reveal the complexities of why agents do not choose sustainable acts from a range of possible acts as a result of being instrumentally incoherent, we will start by discussing Christoph Baumgartner's account. Then we will examine some more recent studies. Baumgartner (2005) discusses relevant empirical psychological and sociological research. His analysis leads to four sets of obstacles that might influence an agent's evaluation of options for sustainable action, which can be summarised as follows.

First, there could be problems with *the observation and valuation of environmentally relevant facts*. Actions always depend on the agent's interpretation of specific situations in specific contexts. Relating the proposed action to the motive of mitigating climate change could involve issues with the interpretation of, for example, greenhouse gas emission data. Should problems arise at this stage, this could block the evaluation process. Once the agent has correctly interpreted the relevant facts, her cognitive patterns could potentially influence the processing of this information, for example in the way she judges environmental risk (see below). Problems with the agent's restricted rationality, for example the wish not to be exposed to "inconvenient truths", can also be grouped under this set of obstacles. Also, the cultural bias in our interpretations and judgements and the question whether agents are able to remove this cultural filter are part of this category of motivational problems.

Second, *conceptual-theoretical limitations* for taking sustainable action may be present. The agent might unreflectively adhere to theories that argue that it is not in her self-interest to choose the sustainable action. One could think here of certain dominant models in economic theory (e.g. the homo-economicus model) prescribing that an agent must act as a "selfish, rational, utility maximizer" (Becker 2006, 18). That is, the homo-economicus model explicitly discourages acting out of altruism or being motivated to take up Humanity's Challenge. Even cooperative action, which aims at contributing to a commonly shared good, such as clean air and water, falls outside of what the homo-economicus theory defines as rational action. When influential people in society publicly apply these economic theories frequently, it can be expected that this will have a limiting effect on the environmentally responsible action of many other people. For persons adhering to the homo-economicus theory, then, it will be difficult to resist free rider strategies.[7]

Third, the agent's *personal limitations* may hamper her deliberation to choose in favour of a sustainable action. As examples of these limitations one could think of competing motives (e.g. when a car is not only used as a means of transportation, but also has the function of giving its owner a certain social status), adherence to environmentally unfriendly values (e.g. freedom of movement, time efficiency, enjoying car driving), and the keeping up of environmentally unfriendly habits that prevent the agent even from considering alternatives for her action (e.g. numerous routine type actions that are being carried out without reflective distance).

Fourth, *valuations and judgements of the agent's responsibility* might hamper the agent's practical reasoning about sustainable action. As agents in the affluent countries hardly experience the ecological crisis (e.g. rising sea levels, as for the inhabitants of Tuvalu, see §1.3), it seems unlikely that they will have responsibility feelings that might influence their deliberation towards taking sustainable action. Feelings of responsibility that stimulate sustainable action can be expected only if the normative reason to help meeting Humanity's Challenge (i.e. the major premise in the PSSA) has become a reason that motivates the agent. In addition, it is unlikely that feelings of guilt for non-sustainable actions will arise when the agent does not experience her action as voluntary, but attributes responsibility for her behaviour to other persons or external circumstances. The criteria according to which judgements about responsibility are commonly formed are different from person to person and relate to the agent's practical self-understanding. Baumgartner argues that the motive of helping to meet Humanity's Challenge becomes action guiding only if the agent aims to belong to a group of significant other people or aims to be a moral person.

Dieter Birnbacher argues for three factors that may hamper an agent's motivation in her deliberation about sustainable action. First, *time preference* means that people prefer present consumption of a good over future consumption. One of the most serious deliberative challenges seems to be that we have a preference for the present merely because of its being present. Another term for this phenomenon is "obliviousness of the future". As an example, Birnbacher refers to the measures taken to protect the Ozone Layer under the Montreal Protocol, which came into force in 1989. The fact that ozone depletion resulted in skin cancer (i.e. it was an immediate problem for people in the eighties) contributed to the relative ease with which this protocol was established. Thus, time preference could have been part of Scheffler's phenomenology of agency (see §1.4 and §4.3) (Birnbacher 2015, 159).

As a second factor, Birnbacher mentions *uncertainty* about the actual impact of climate change on the quality of life of people in the affluent countries. Although there is little room for doubt about the existence of anthropogenic climate change (see §1.3), there is uncertainty about who will be the victims of global warming. Since people's decision about mitigating actions depends on calculations of the consequences for themselves and for their immediate descendants, inaction is likely to occur as a result of uncertainty about who will be affected. Another uncertainty is whether one's sustainable action will make any meaningful contribution to mitigating climate change. A successful strategy for the taking up of Humanity's Challenge depends on specific feedback about their contribution being given to those who take sustainable action. Birnbacher argues that our control beliefs have to be strong in order for our motivation to be stable enough. This means that agents need frequent, good-quality feedback about the effectiveness of their sustainable action. Another uncertainty is that we are not sure whether the generations after us will continue taking sustainable action with the same goal of meeting Humanity's Challenge as we have. We cannot expect that later generations will cooperate simply because our generation has taken the lead (Birnbacher 2015, 160).

The third factor that influences us while deliberating is *limited altruism*. In Birnbacher's view it seems unlikely that people will choose in favour of sustainable action if they rely on the moral motivation to do good to future people. Future people as beneficiaries of one's action are spatially and temporally distant. Thus, according to Birnbacher, since the potential victims of the problem of climate change remain unidentified, abstract, and anonymous, current people in the affluent countries are not encouraged to become motivated to act sustainably (Birnbacher 2015, 161).

In relation to Baumgartner's first category of deliberative problems, the environmental psychologists Gisela Böhm and Carmen Tanner argue for the importance of *risk assessment* when assessing options for taking environmentally friendly action. Environmental risks differ from other risks in three ways. First, they are characterised by high complexity and uncertainty, entailing complicated causal relationships and multiple consequences (e.g. the risk of acidification of oceans caused by anthropogenic $CO_2$ emissions, and the risk of destruction of human habitat due to flooding). Second, environmental risks often emerge from aggregated behaviours of many individuals (e.g. use of fossil fuels) rather than from a single activity. Third, the consequences of environmental hazards are often temporally delayed and geographically distant (e.g. historic greenhouse gas emissions in the affluent countries that cause people in developing countries to become climate refugees) (Böhm and Tanner 2013, 16).

Risk perception refers to people's subjective judgement about the risk that is associated with some action. Böhm and Tanner report about phenomena that are observed when people are asked for their subjective risk assessment. For example, people's judgement is influenced by the ease with which they can bring to mind examples of an event: the easier this is for them, the more likely they are to overestimate its occurrence. Therefore, when the media frequently report about car accidents, we are more likely to think that car driving is risky. Or, people have a tendency to believe that they are more likely to experience positive events and less likely to experience negative events than similar other people. Such unrealistic optimism can have important implications in that people fail to take preventive measures when they perceive no personal risk (Böhm and Tanner 2013, 18).

In a recent meta-study, the environmental psychologist Robert Gifford provides an overview of psychological barriers that stand in the way of the agent's deliberation in favour of taking sustainable action. Similar to our key question, Gifford asks: "If so many people are concerned about climate change, the environment, and sustainability, why are more of us not doing what is necessary to ameliorate the problems?" (Gifford 2011, 290). Gifford arranges no fewer than 29 of such psychological barriers into seven categories. I will mention a few of the categories that seem particularly relevant to the deliberative phase. First, people may face *cognitive limitations* (Gifford 2011, 291). For example, one is ignorant about the problem of climate change, or one is numb regarding the environment. The latter barrier occurs, for example, if one hears too often about environmental problems. A second category of psychological barriers is *ideologies*. For example, a belief in free enterprise capitalism, based in turn on the belief in the freedom of the commons, has "led to the devastation of fisheries, forests and landscapes around the world", and thus influences deliberation about sustainable action (Gifford 2011, 293).

A third group concerns people's tendency to make *comparisons with other people*. Agents tend to compare their actions routinely with those of others and derive what they ought to do from their observations about what is the "proper" course of action. Likewise, agents ask: "Why should I change if they won't change?" Usually, well-known persons or other nations are cited as not cooperating, which serves as a justification for inaction (Gifford 2011, 294). Fourth, like Böhm and Tanner, Gifford mentions *perceived risk* as a psychological barrier. When people are deliberating about changing their behaviour, they may perceive various forms of risks that could hold them back, leading to questions like: Will the new plug-in electric vehicle work, given that the batteries contain new technology? (*functional risk*) Suppose that one has plans to move in a few years' time; will this match with the payback period of the solar panels? (*financial risk*) How will significant others judge my changed behaviour? (*social risk*) (Gifford 2011, 296).

A final study worth noting in connection with the deliberative phase considers "the relationship between environmental self-identity, obligation-based intrinsic motivation" and taking sustainable action. The environmental psychologists Ellen van der Werff, Linda Steg, and Kees Keizer define environmental self-identity as "the extent to which one sees oneself as a type of person whose actions are environmentally friendly" (Van der Werff *et al.* 2013, 1258). The authors argue that some people act sustainably without any external incentives to do so, but rather because they have an intrinsic motivation, which is obligation-based – that is, "being motivated out of a sense of obligation related to approval from oneself or others" (Van der Werff *et al.* 2013, 1259). The research by Van der Werff *et al.* indicates that the intrinsic motivation of people who take sustainable action results from their strong environmental self-identity. More precisely, the environmental self-identity of agents is related to their obligation-based intrinsic motivation, which in turn promotes taking sustainable action (Van der Werff *et al.* 2013, 1264). From this study it can be concluded that if agents face a lack of external rewards as an obstacle to their deliberation in favour of a sustainable action, then it may be more effective to strengthen the agent's environmental self-identity, than to work on improving the external incentives.

The above examples of deliberative problems are not intended to be an exhaustive list. The studies discussed above are meant to show the variety of motivational problems that can occur when agents deliberate about proposals for action, including sustainable actions. In the remainder of this chapter we deal with the two remaining philosophical questions, identified in §1.3, that influence an agent's deliberation about proposals for action.

## 4.3 Sustainable action and common-sense morality

In this and the next section, we address the questions: If the agent reasons from common-sense morality, then what can be expected of her when it comes to acting in conformity with the norm to meet Humanity's Challenge?

And, what can be said about the motivating force of the obligations on the agent to respect the rights of future generations? Whereas the former question, quite obviously, relates to the ethical theory of common-sense morality, the latter question is connected with the metaphysics of current and future people. Both common-sense morality and positions from the metaphysics of future persons could provide agents with excuses not to act in a way that contributes to efforts to meet Humanity's Challenge.

We will not, however, enter into the debate as to whether common-sense morality and certain metaphysical theories of future persons are true or untrue. In fact, as we will see, both common-sense morality and the theory of future people (i.e. Parfit's Reductionist View) have been criticised. Instead of taking a position in that debate, I will show that these theories facilitate agents in self-indulgently reconstruing their situation in such a way that they do not have to act in accordance with the major premise of the PSSA, which they have adopted (see §4.2). In other words, their self-indulgent reconstruction hampers these agents in their deliberation about taking sustainable action. Since we are considering the agent's motivation to reduce her luxury emissions, I will argue that agents ought to resist the temptation to be lenient with themselves that these theories create. However, as we will see, institutional help will be required to support agents in resisting these temptations.

In the rest of this section we discuss the first of the two questions. My claim is that reasoning from the point of view of common-sense morality may form an impediment to the agent's motivation in the deliberative phase that is specific to the taking of a sustainable action. That is, if my motive to meet Humanity's Challenge pushes me one way, and my motive to follow common sense pushes me another, then where will my overall orientation lead me? In order to argue that there is a specific problem for sustainable action here, I will first provide some of the main claims of common-sense morality. Thereafter, I will continue with the analysis of the impediments that follow from adherence to common-sense morality in the deliberative phase.

Common-sense morality allows agents to give weight to their personal interests with regard to ethical theories by establishing a personal point of view as opposed to the moral point of view. Most agents take it as self-evident that there are certain significant other people to whom they have special obligations. These are people to whom they stand in certain relations, such as children, parents, friends, pupils, patients, clients, and colleagues. Agents believe that they ought to try to give these people certain kinds of benefit and prevent them from suffering harm (Parfit 1987, 95). Seen in that light, common-sense morality allows agents discretion to treat significant others differently from what morality prescribes. Common-sense morality seems to be the default moral position of agents, which they generally tend not to analyse and discuss (Van den Hoven 2006, 25).

According to common-sense morality, agents ought to have room to pursue the projects, plans, and commitments that are most dear to them for three reasons (Van den Hoven 2006, 125). First, agents realise themselves as autonomous persons through so-called ground projects. That is, people see it as an expression of themselves as autonomous persons to make their own choices with regard to the projects they are committed to (Van den Hoven 2006, 138). Second, ground projects are crucial for constituting a person's integrity: giving them up would make it difficult to realise a unity in one's personal life (Van den Hoven 2006, 140). Otherwise put: life would simply not be worthwhile in the absence of ground projects, with which one can identify; in fact, one would not even have a reason for going on (Calhoun 1995, 241). Third, agents take care of those to whom they stand in a significant relationship. According to common-sense morality, ethical theories must take seriously the commitments that agents have taken on in the form of the care of their loved ones. Caring for others not only forms an inalienable part of human life, it also has an identity-conferring element: "The persons I care for and those with whom I stand in specific relationships determine who I am and how I conceive the world around me" (Van den Hoven 2006, 143).

Shelly Kagan construes common-sense morality as supporting the claim

> that there is a limit to what morality requires of us. Many acts which would lead to the best results overall nonetheless are *not* required of us by ordinary morality, typically because the sacrifice would be too great to demand it of us. I am not required to devote my free time to fighting political oppression, nor must I give up my luxuries to support cancer research.
>
> (Kagan 1989, 2, emphasis Kagan)

Common-sense morality, thus, is of importance in the philosophical debate about the limits of morality. Central to this debate is the question of whether ethical theories (e.g. a demand to mitigate climate change) can lead to unreasonable demands on agents (e.g. to eat less meat, to install solar panels, to reduce air travel). Common-sense morality allows agents to favour their personal interests, even if by doing so they fail to live up to, for example, the moral requirement to help meet Humanity's Challenge.

A first feature of common-sense morality, Kagan argues, is that it gives agents the *option* of performing (or not performing) an act that is morally required, and that it supports the belief in the existence of such agent-centred options. This is not to say that no sacrifices at all are required by common-sense morality. However, these sacrifices tend to be rather modest and limited. Agents are free to make sacrifices to promote the overall good; they are not, however, required to make them (Kagan 1989, 3).

A second feature of common-sense morality is that "it lays down certain strict limits on our actions – forbidding various types of acts *even* if the best consequences overall could be achieved only by performing such an act" (Kagan 1989, 4, emphasis Kagan). Restrictions on our actions arise, for example, from rights, and from prohibitions against intending harm. Other restrictions may arise from special obligations the agent has taken on, for example past promises made to significant others or duties related to one's professional role. Common-sense morality forbids violating such special obligations, even if this is the only way to achieve the best consequences overall. Thus, common-sense morality endorses the existence of *constraints* – that is, an agent may be prevented from executing a morally required action because of the restrictions or special obligations mentioned above (Kagan 1989, 4).

Kagan confronts common-sense morality with the view that "the demands of morality pervade every aspect and moment of our lives – and we all fail to meet its standards". This is what he calls the extremist's position (Kagan 1989, 2). The extremist's position involves taking a hard look at one's life plans and reshaping them according to what morality requires. At the very least it involves taking into account the long-term effects of one's actions and not only the more immediate ones (Kagan 1989, 8). This means that once the agent's special obligations are met, she is required to devote herself to making her greatest possible contribution to the overall good (Kagan 1989, 9). The extremist's position is that there is no limit to what one might be called upon to sacrifice in the pursuit of the good. Common-sense morality, on the other hand, claims that there must be a limit to the costs that morality can inflict upon an agent. Common-sense morality seeks a defence of options in an appeal to considerations of the costs, to the agent, which would be imposed by various moral demands (Kagan 1989, 21).

In sum, common-sense morality tempers the relentless pursuit of the overall good. Options give the agent permission to pursue her personal interests rather than the overall good. Constraints forbid certain courses of action, even when they are the necessary means to achieving the overall good. Within the limits of the constraints, there are occasional demands, but typically only rather modest demands are made.

Notice that we focus the problem of mitigating climate change on the reduction of *luxury* emissions by people in the *affluent* countries. When taking action to work on Humanity's Challenge, agents in affluent countries need to scrutinise their common-sense moral points of view in order to reconcile their existing interests, projects, and attachments with the obligation to reduce their luxury emissions that they owe to future people. If there were an a priori demarcation between the personal and the moral point of view, then chances would be very low that a commonly shared view would be found on what to do to meet Humanity's Challenge. The question, thus, is whether the normativity of meeting Humanity's Challenge and the perhaps rather weak motive derived from it (see §1.4) can compete with the agent's motivation to care about her existing interests, projects, and attachments.

By referring to their common-sense personal points of view, agents in affluent countries feel supported in excusing themselves from taking sustainable action, on the grounds that the moral point of view of taking up Humanity's Challenge may demand too much from them, even if only the reduction of their luxury emissions is at stake. For example: common-sense morality facilitates agents who, in their deliberation about options for action, argue that it is over-demanding to ask them to give up a holiday flight to Thailand and instead take the train to France. By elaborating how common-sense arguments enter the deliberative phase, I argue that our *direct* motive for sustainable action will not be sufficient to strike the balance between this form of morality and endeavouring to mitigate climate change. In my view, it will be very unlikely that individual agents by themselves will be able to enlarge the horizon of their common-sense moral position to the extent that they start endeavouring to meet Humanity's Challenge. Therefore, having considered the psychological tendencies that dominate the agent's deliberation about taking a sustainable action (see §4.2), I conclude that institutional support will be required for promoting this sort of action (see §7.3). Moreover, in Chapter 9, I will develop an *indirect* motive (see §2.4) for agents to make a commitment to help meeting Humanity's Challenge.

We now turn to analyse the agent in an affluent country and the impediments she faces from adherence to common-sense morality while she is deliberating about possibilities for taking actions that contribute to meeting Humanity's Challenge. By this we mean actions that aim to decrease her luxury emissions. Of course, it is unreasonable to demand that a poor mother in Sub-Saharan Africa, who is struggling to get food for her children, donate to the people of the islands of Tuvalu, who in the near future might lose their land as a result of rising sea levels. For agents in circumstances such as these, common-sense morality rightly asks us to respect limits as to what ethical theories can appropriately and acceptably demand. However, to what extent can the agent in an affluent country, while deliberating about proposed actions to reduce her luxury emissions, demand similarly that we respect her personal point of view as opposed to deferring to the moral point of view?

Let us consider again the case of S, who proposes that she and her fellow homeowners place solar panels on their roofs (see §1.1). It appears that it accords with S's understanding of common-sense morality to take the relatively small burden of investing €5,000 and to receive this amount back, roughly within eight years, by means of a reduced monthly energy bill. The other homeowners, however, consider losing the possibility to invest their money elsewhere as a "cost to agent" that in their view is too high, even though they are aware that they do not need to refrain from any other purchase that might be needed to ensure they continue living above the threshold of wealth they deem appropriate. It seems, then, that S is of the opinion that she has valid reasons to propose installing the solar panels – that is, she believes the PSSA to apply to all the homeowners, whereas the other homeowners believe they have the option to take this action or not.

This example shows that from the common-sense moral point of view it could be argued that Humanity's Challenge is demanding too much from us, even for agents in affluent countries who have accepted the normative reason for human beings collectively to limit luxury emissions. We seem to have reason to believe that we might run into moral dilemmas in the choice between striving to meet the challenge (e.g. by accepting certain restrictions on our actions) and our direct personal interests (e.g. the care of our loved ones). The agent might, then, be tempted to limit the burden on herself that she would feel if she were to confront her significant others with the consequences of engaging in sustainable action, which she believes could lead to a reduction in their well-being. In summary, common-sense morality justifies agents' claim that limiting their luxury emissions is over-demanding, given that they have duties towards significant others.

In the conceptual-phenomenological analysis, we consider the agent's experience of these situations in order to try to understand her deliberation about the options for action at her disposal. We will start the analysis of the agent's perspective by considering the way Wouter Peeters *et al.* have discussed it.

These authors argue that Scheffler's phenomenology of agency (see §1.4) supports certain psychological tendencies that constitute potential obstacles to the agent's motivation to act sustainably. First, there is an asymmetry between our experiences of our acts, and those of our omissions. Whereas we ordinarily experience our acts as acts, we experience our omissions as omissions only in specific contexts. This feature seems to support an attitude among agents that involves feeling a distinct responsibility for what they themselves do, but not for what they merely fail to prevent (Peeters *et al.* 2015, 59). The relevance for the deliberative phase is that agents seem to be less aware of the act of changing their behaviour in order to reduce their luxury emissions than they are of the act of continuing with these emissions.

For example, refraining from taking the drive in a sports utility vehicle for fun on a sunny Sunday afternoon (see §1.2) appears less significant in the agent's experience than going for the drive. If agents consider not going for the drive, they need an effective motivational strategy to help them overcome this anticipated gap in their experience. It seems that they could compensate for the phenomenological insignificance of the act of refraining by going for a bicycle ride on that same sunny Sunday afternoon. The experience of the cycling would replace the experience of the car drive, and the agent would not face the experiential insignificance of refraining from the car drive. From the first-person perspective of the agent it thus seems easier to replace the drive by a bicycle ride than to just refrain from it and stay at home without having a different pleasant experience.

Second, we tend to experience our causal influence as inversely related to spatial and temporal distance. This seems to support the psychological tendency for individual agents to feel a special responsibility towards those to whom they stand in a significant relationship, rather than to strangers (Peeters *et al.* 2015, 73). This phenomenon pops up in the deliberative phase for agents in the affluent countries, since the problem of climate change is remote to them. People in the highly industrialised countries experience hardly anything related to climate change. As a consequence, when deliberating about the option to act sustainably, agents fall back upon their immediate surroundings, including their relations to significant others, as their primary area of responsibility, without *experiencing* their moral failure.

To continue with the example of the non-sustainable drive in the SUV, the effect of this drive on the climate is difficult to experience, since it seems imperceptibly small and remote. When deliberating, the agent faces difficulties in weighing the experience of the climate effect against the experience of the joyful drive in an SUV. However, if the agent feels a responsibility to significant others to provide a joyful afternoon by going for the drive, she could argue that making the bicycle ride as joyful as the car drive might be an effective motivation for her loved ones. Of course, the bicycle ride cannot fully replace the experience of the car drive. However, the agent's deliberative choice in favour of cycling instead of driving could be crucial in changing the perception of cycling as an act that can generate as much pleasure as driving. Though it may not be sufficient, the agent's all-things-considered judgement to go on the bicycle ride instead of the car drive seems of crucial importance to pursuing the sustainable action.

Third, when an outcome is the result of the joint actions of a number of people, including ourselves, we tend to see the contribution of our own agency to a much lesser extent than we do when we view an effect as having resulted solely from our own actions. This feature supports the common-sense practice of treating the individual agent as primary locus of responsibility (Peeters *et al.* 2015, 91). In other words, agents are held responsible for outcomes solely resulting from their own actions, as opposed to outcomes that are the joint result of the actions of a number of people (Peeters *et al.* 2015, 24). Indeed, our method of investigation (see §2.4) has been set up in line with this common-sense view – that is, our method is to focus in the first instance on an understanding of the problem with motivation for sustainable action for individual agents (see §2.2).

However, the analysis of the agent's first-person perspective in the deliberative phase makes clear that to overcome these psychological tendencies, supported as they are by common-sense morality, institutional interventions are needed. Meeting Humanity's Challenge is a collective action problem, in which the first-person temptation to free ride is powerful (see §4.2). The agent's deliberation about options for action reveals that we cannot assume that she will be able, by herself, to enlarge her moral horizon so as to include future generations.

The first feature of Scheffler's phenomenology makes clear that agents in their deliberation need help to take into account that changing their behaviour in order to reduce their luxury emissions is as significant as the act of continuing with these emissions. If we take the example of the wasteful drive a step further, it will be *understandable* from the agent's point of view that it is instrumentally coherent to promote a bicycle ride as an alternative to the car drive. However, institutional help will be required for agents to deal with the difference in *experience* when cycling is chosen as an alternative to car driving. Somehow, in her deliberation about action proposals, the alternative experience of cycling must be brought to the agent's attention. As we cannot rely on the agent herself in this regard, it seems to me that institutional interventions are required. It is, of course, another question *how* the communication of alternatives to wasteful driving on Sunday afternoons can be done effectively. This is the competence of disciplines such as marketing and communication, which I leave out of the scope of this book.

Scheffler's second feature seems even more difficult for the individual agent to overcome during the deliberative phase. Considering another experience while deliberating cannot bridge the phenomenological distance between the agent's experience of the car drive and the effect of the drive on the environment. Reasoning about refraining from the wasteful drive requires strong imaginative and moral qualities, since the distance of the drive's detrimental effects on the climate in space and time has to be bridged. To assist agents in acquiring the deliberative abilities to cope with this psychological difficulty, it would seem helpful to provide scientific information describing the problems current people in developing countries face with rising sea levels and relating those phenomena to their causes in terms of historical emissions. This would ensure the agents possess, while they deliberate, knowledge of the effects of wasteful luxury emissions.

However, although necessary, these interventions will not be sufficient to ensure individual agents decide in favour of sustainable actions. Institutional help will be needed to underline the normative aspects of providing aid to people suffering from harm. This could be done, for example, by communicating a principle such as the one advocated by Peter Singer: "If it is in our power to prevent something very bad happening, without thereby sacrificing anything of comparable moral significance, we ought to do it" (Singer 2011, 199). In addition, there seems to be room to look critically at the scope of common-sense morality, specifically where taking up Humanity's Challenge is concerned. When agents in affluent countries defend their luxury emissions with reference to common-sense moral considerations, I consider myself firmly in the camp of Shelly Kagan, who states about our common-sense beliefs:

> Ordinary morality judges our lives morally acceptable as long as we meet its fairly modest demands. It is not surprising that this view should be so widely – and uncritically – held: it is not pleasant to admit our failure to live up to the demands of morality. But the truth remains that we are morally required to promote the good and yet we do not. Faced with this realization what we must do is change: change our beliefs, our actions, and our interests. What we must not do – is deny our failure.
>
> (Kagan 1989, 403)

Our analysis of the potential impact of common-sense morality in the deliberative phase indicates that the inability of agents to vividly imagine future people as potential victims of their current non-sustainable behaviour constitutes an intractable problem for meeting Humanity's Challenge. Agents seem to have great difficulty with feeling the motivational force of abstract concerns about future generations. The persistence of this problem seems specifically relevant to sustainable actions. Although the credit restructuring officer (see §3.6) deals with morally dilemmatic situations, these are situated entirely within her moral horizon. Facilitated by the mass media, the doctor without borders (see §3.3) has been made aware of the people who need emergency aid. These institutions helped him enlarge his moral horizon, to the extent that he even decided to change his professional life drastically. When people are standing in front of us, when we look into their faces, we seem to have less difficulty with including them in our moral deliberation than we have with people who will live in the distant future (Birnbacher 2009).

We can conclude, therefore, from the relatively simple example of refraining from a wasteful drive on a sunny Sunday afternoon and replacing the drive with a bicycle ride, that the widening of people's moral horizon, with the goal of promoting sustainable action, is necessary to remove a hampering factor during the deliberative phase. Considering the issues that individual agents face, it seems very unlikely that they will have the strength of will to enlarge their moral horizon without any institutional support. To cope with the temptation to be self-indulgent and to deliberate in favour of not taking a sustainable action, as facilitated by common-sense morality, I argue that we need to strengthen the motivation of agents in – at least – two ways. First, we need to look for institutional interventions. In Chapter 7, therefore, I will elaborate proposals to this end. Second, in addition to the direct motive for sustainable action, an *indirect* motive is needed that makes us change our behaviour towards sustainable behaviour (see §2.4). In Chapter 9, I will propose an indirect motive that follows from a practical understanding of ourselves when we follow a strategy of non-sustainable behaviour. This enriched practical self-understanding that comes from understanding our motivational weaknesses when taking sustainable action, like the norm of meeting Humanity's Challenge, again has normative force on us.

## 4.4 Sustainable action and practical self-understanding

There remains the fourth and last philosophical question from §1.3 that we have to address: What can be said about the motivating force of the obligations to respect the rights of future generations? I argue that underlying this question there is a debate about the metaphysics of current and future people that potentially prevents the agent from reaching a judgement about which action to take, during her deliberations about the minor premise in the PSSA. If the agent holds certain beliefs about the metaphysics of future people, then this may hinder her from judging in favour of a sustainable action. That is, the agent's belief about the metaphysics of future people may override her motive for taking up Humanity's Challenge. Agents may even return to questioning the major premise of the PSSA when their metaphysics of future people is incompatible with the understanding of themselves that they have to presuppose when they act in social contexts. The problem I discuss is the deliberative problem that arises as a result of holding two incompatible conceptions of the human being: on the one hand a normative-phenomenological conception that we must presuppose if we want to understand our agency as *current* people in social institutions, and, on the other hand, a reductionist conception that we may apply simultaneously to *future* people. I will now explain how this incompatibility leads to a hampered motivation during the deliberative phase.

As we have analysed so far, there are several psychological obstacles that may have an influence on the agent's instrumental rationality while deliberating about taking sustainable action (see §4.2). Furthermore, we have shown that the agent's deliberation is under the influence of features of commonsense morality. Without the support of an institutional policy, such influence may lead to indecisiveness or the continuation of non-sustainable behaviour (see §4.3). While we have two phases of the Rubicon model still to analyse (see Chapters 5 and 6), and the problem formulated in this book (see §1.5) still to answer, I take it that deciding about sustainable action can become a burden for agents (i.e. "Why is it so hard to engage in sustainable action?") In other words, the agent has accepted the normative reason to contribute to efforts to meet Humanity's Challenge and, following the PSSA (see §4.2), believes that, through executing an action with certain characteristics, meeting Humanity's Challenge will be realised or its realisation will be furthered; she, however, refuses to reach the all-things-considered judgement that this is the action to be taken.

In their discussion of strategies that agents adopt when they are practically deliberating, Cass Sunstein and Edna Ullmann-Margalit emphasise delegation as a familiar way of handling decisional burdens. "People might, for example, rely on a spouse or a friend or choose an institutional arrangement by which certain decisions are made by authorities established at the time or well in advance" (Sunstein and Ullmann-Margalit 1999, 10). In an effort to reduce the agent's burdens both before and at the time of reaching the all-things-considered judgement, delegation is a strategy that exports the burdens of reaching the judgement to someone else: "A typical case involves an agent who seeks to avoid responsibility ... and who identifies an available delegate whom he trusts to make a good, right, or expert decision" (Sunstein and Ullmann-Margalit 1999, 16). For our conceptual-phenomenological analysis of deliberation about sustainable action it is crucial, however, to consider the specific case of intrapersonal delegation. This happens when the agent delegates her decision about a sustainable action to her future self. I refer here to the situation in which agents procrastinate as a result of the delegation of the decision to their future selves (Sunstein and Ullmann-Margalit 1999, 20). The result is a motivational obstacle in the sense that, despite her acceptance of Humanity's Challenge as a normative reason for the collective of human agents, the agent stops deliberating, because she understands herself to have delegated the decision to her future self.

Our focus in this section is on setting out this reason for not reaching an all-things-considered judgement. As I will elaborate, the reason is that the agent believes that the identity of future people is fully contingent upon the actions of current people. Due to holding two incompatible conceptions of human nature, the agent could run into a deliberative problem that blocks her motivation to act should she conclude: The identity of future people is formed through entirely contingent events; if they are born, they will owe their very existence to the actions that we will have taken. Since our actions constitute their identity, they will not regret what we have done. As changing my behaviour towards more sustainable behaviour is a burden for me, I am justified in delegating a decision about taking sustainable action to my future self. Hence, I don't need to act sustainably now.

We will consider this problem as follows. We start with an overview of one of the most influential reductionist conceptions of human nature, which supports delegation of decisions to an agent's future self: Parfit's Reductionist View. This metaphysical conception of the human being (Shoemaker 2015, §3) has dominated the debate about intergenerational justice for some time. I begin the discussion of the Reductionist View by including Christine Korsgaard's response to Parfit. Then, as an alternative to Parfit's view, I outline the normative-phenomenological conception of agency according to which current people can understand the human practice of social cooperation. Subsequently, I problematise the Reductionist View by arguing that agents ought to resist the self-indulgent temptation not to resolve the above-mentioned incompatibility, but rather to procrastinate about taking sustainable action through delegation to their future self. As stated earlier (see §4.3), it is not my aim to be definitive about the question whether the Reductionist View is true. However, I do want to emphasise the motivational problems that agents encounter when this view enables them to put off sustainable action.

The basis for Parfit's account of personal identity is the *Psychological Criterion*, which tells us that a psychological connection exists when a psychological state at $t_2$ is causally related in some way to a psychological state at $t_1$:

> X at $t_1$ is the same person as Y at $t_2$ if and only if X is uniquely psychologically continuous with Y, where psychological continuity consists in overlapping chains of strong psychological connectedness, itself consisting in significant numbers of direct psychological connections like memories, intentions, beliefs/goals/desires, and similarity of character.
>
> (Shoemaker 2015, §2.1)

According to Parfit, the psychological criterion is *reductionist*, since a person's identity over time consists in the holding of certain particular facts, which can be described in an impersonal way – that is, without presupposing the identity of the person, or even without explicitly claiming that the person exists. On the Reductionist View, each person's existence simply involves the existence of a brain and body, the doing of certain deeds, the thinking of certain thoughts, the occurrence of certain experiences, and so on (Parfit 1987, 211). Note that Parfit distinguishes carefully between a person being a separately existing entity, which he rejects, and a person being "an entity that is *distinct* from a brain and a body, and such a series of [interrelated physical and mental] events" (ibid., emphasis Parfit).

The other feature of personal identity, according to Parfit, tells us about the importance of personal identity, namely that it is not important at all. Parfit rejects the claim that a person, considered as the subject of experiences, can explain psychological unity through her ownership of experiences (Parfit 1987, 214). Personal identity is not what matters; what matters is *Relation R*: "Psychological connectedness and/or continuity with the right kind of cause. In an account of what matters, the right kind of cause could be any cause" (Parfit 1987, 215). In order to argue for the non-importance of personal identity, Parfit starts by stating that the Psychological Criterion is the most plausible reductionist criterion. The criterion entails that, in order for X to be identical to Y, they must be uniquely psychologically continuous. However, psychological continuity is potentially a one-to-many relation – that is, it could hold between me-now and more than one person in the future. But identity is an equivalence relation; it holds only one-to-one. Parfit, therefore, needs to include a "no branching" clause in the Psychological Criterion to avoid a contradiction (Shoemaker 2015, §2.4).

Parfit continues to argue that Relation R matters instead of personal identity by considering the case of fission (Parfit 1987, 254–256). To test the view that personal identity matters (the Cartesian Ego View), he asks what would happen if his brain were divided over two of his brothers, who are his identical triplets. What will happen to Parfit after the surgery? Which of the brothers will be Parfit? Parfit analyses these four cases: (1) he does not survive; (2) he survives as one of his brothers; (3) he survives as the other; (4) he survives as both. On the basis of an assumption about the successes of brain surgery, Parfit argues that the first three options have to be dismissed immediately. The fourth case seems the most complicated one. However, after a lengthy argument, this case also cannot be sustained. There will be two people, each of whom will have the body of one of Parfit's brothers and will be fully psychologically continuous with Parfit, because each has half of his brain (Parfit 1987, 260). Parfit concludes that the case of fission is not about his survival, but about psychological continuity, on which the Reductionist View is based. Thus, personal identity is not what matters.

Since on the Reductionist View "personal identity cannot be defensibly defined in terms of the unity of a subject of experiences" (Schechtman 2005, 6), the main consequence of this view, which is relevant to deliberation about sustainable action, is that it allows delegation to future selves. As I have already stated, delegation to future selves causes agents to postpone their decision about sustainable action. Therefore, I will now begin problematising the Reductionist View by discussing Korsgaard's response to Parfit. In particular, note that it is a standard position in practical philosophy that "persons should be viewed primarily as agents, and the unity of persons as the unity of agency" (Schechtman 2005, 14). Korsgaard's account relies on such an agency-based view of personal identity.

Korsgaard begins her discussion of the Reductionist View by stating that activity and passivity are aspects and not parts of human nature, and that each can be reduced to a form of the other. She argues, however, "that from a moral point of view it is important not to reduce agency to a mere form of experience" (Korsgaard 1996b, 364). Korsgaard's point is that from a first-person point of view, we have reasons for regarding ourselves as the same rational agent as the one who will occupy our body in the future (Korsgaard 1996b, 369). To arrive at this, first she considers the problem of identity at any given time. As agents we need to align our various motives with each other; we must eliminate conflicts between them. We are a unified person at any given time, because we must act, and we have only one body with which to act. Another element of our pragmatic unity follows from the standpoint from which we deliberate and choose. It is I, who must choose among conflicting desires which one to act on. We have to integrate the actions we take into the narrative of our lives with which we identify. This does not require a separately existing unity or involve a deep metaphysical fact. Instead, it is a practical necessity imposed upon us by the nature of the deliberative standpoint (Korsgaard 1996b, 370).

Second, Korsgaard considers the unity of agency over time. As she argues, things we do, such as choosing careers, pursuing friendships and family lives, are intelligible only in the context of projects that extend over long periods (Korsgaard 1996b, 371). In particular, endeavouring to meet Humanity's Challenge presupposes the continuity of our identity and agency. The normative-phenomenological conception of agency that I propose below gives us the possibility to derive reasons for action from a commitment we have made far into the future. That is, from a practical point of view, we must consider ourselves as unified agents over time.

In setting up the method of this book, we have elaborated the relevance of a practical, first-person understanding of our motivation to act sustainably (see §2.4). Understanding why agents act as they do requires an understanding of what moves agents, while they are engaged in practical deliberation, commitment, and action execution. The problem with the Reductionist View is that it is a third-person understanding of personal identity that does not help us with the first-person understanding of our agency. As Korsgaard's argument shows, it does not seem possible to integrate the Reductionist View with a view that makes sense from the practical deliberative standpoint. As she states: "We think of living our lives, and even of having our experiences, as something that we *do*. And it is this important feature of our sense of identity that Parfit's account leaves out" (Korsgaard 1996b, 378).

Any responsible policy for mitigating climate change requires that at some moment in the not too distant future global emissions of greenhouse gases should peak and from there on start decreasing substantially for at least a couple of decades. This global course of action to reduce greenhouse gas emissions must continue until the desired levels are reached (see §2.2). This means that if we want to meet Humanity's Challenge, we need to understand ourselves as agents in terms of a conception of agency that enables us to view our actions in relation to our commitment over a long period of time. In order to take sustainable action over this long period, we need a conception of agency that is *normative* in that it provides practical guidance for agents. This conception thus must be constitutive of our agency (Harnacke 2015, 62). For example, the conception we are looking for must allow for a robust assignment of responsibility and desert. If a conception of agency aims to provide practical guidance for agents, then it must not lead us to evade our responsibilities through facilitating us in claiming loose psychological connections to our actions. At the same time our conception must be *phenomenological* in that it should be based on the way we experience ourselves as agents. That is, the conception must tell us something about our agency that we accept as something we know from experiencing our agency.

I have taken as a methodological assumption a minimally individualistic concept of institutions (see §2.2). Note that in the following paragraphs I restrict the concept of institutions to social institutions, in other words organisations. Institutions are composed of human individuals, who are in an ongoing discourse with each other about the strategies, policies, and actions they are pursuing as representatives of the institution. Agents pursue those strategies, policies, and actions with the view of furthering the goal of the institution. What the institution in fact represents is intersubjectively shared among the representatives through an ongoing discourse between them. All actions that in common-sense language are ascribed to the institution are in fact carried out by representatives. Agents work according to commonly shared role descriptions, or at the very least a personal interpretation of such descriptions to which the representative can be held accountable in the light of the purpose of the institution.

The understanding of institutions as being composed of human individuals implies that agents ascribe to each other a form of autonomy. Such autonomy is based on two characteristics. First, it follows from the concept of an institution that agents work to realise its purpose, at least as they understand it. Agents expect one another to contribute to achieving the commonly shared goal. They expect one another to make a contribution within the context provided by the institution's features, for example its mission, strategy, structure, systems, and culture. Consequently, agents presuppose the skills necessary to perform their role; at least they presuppose that those skills can be learnt.

What are basic agency skills? Institutions are founded on the premise that agents can act through practical deliberation, committing themselves to a plan, executing the plan, and evaluating what they have achieved. In other words, agents are expected to be capable of following the phases of the

Rubicon model. Deliberation, commitment, execution, and evaluation are necessary in all roles and all tasks within an institution. We thus necessarily attribute a certain form of autonomy to representatives of an institution in carrying out their role or task. Without this form of autonomy, cooperative agency in institutions seems inconceivable.

Second, being an agent in an institution is not a matter that remains fully external to the person. The agent receives external goods from the institution (e.g. salary, position, work space) and, more importantly, develops over time the internal goods that allow her to increase her excellence within the practice of the institution. Obviously, it could be objected that the possibilities that agents in institutions have to develop the internal goods that contribute to their conception of a good life, and of being autonomous agents, depend on the culture of the institution. In reply, I note that I have chosen a concept of voluntary and purposive action (see §1.2), and that agents are situated in affluent countries (see §1.5). Consequently, even in those cases in which agency in the institution is entered into reluctantly, I take it that the agent's reason for doing so still contributes to her ideal of a good life. That is, I take it that agents have voluntarily decided to join such an institution in order to perform their agency.

One's capabilities in terms of contributing to the institution's purpose are gradually extended, for example one becomes a better, more experienced machine worker, musician, or sales manager. Being a member of an institution thus delivers the internal good of contributing to one's ideal of a good life. That is, in our ideal of a good life our agency in institutions, somehow, plays a role. Hence, although there can be numerous situations in which agency in institutions will be far from perfect, we must acknowledge that it enhances our capacity to develop and pursue our ideal of a good life.

From our methodological standpoint it follows that we take the agent's experience as the object of study. The basic phenomenon we consider is the agent, who experiences her acting. Although agents live through a number of different experiences, the experiencing itself remains a constant with regard to the agent whose experience it is. Agents experience their self-identity through time. As we understand it, the self is "the sum total of its narratives, and includes within itself all of the equivocations, contradictions, struggles and hidden messages that find expression in personal life" (Gallagher 2000, 20). For our conception of agency, this means that we do not need to posit the self as a separate entity over and above the stream of consciousness. We consider agents to be capable of integrating a sense of ownership of their actions with their aims, ideals, and aspirations. Thus, agents deliberate, plan, and execute actions for which they take responsibility (Gallagher and Zahavi 2015, §1).

Mitigating climate change means that we commit ourselves to taking sustainable action as though we were going to act towards that end over decades and centuries. Even if fulfilling Humanity's Challenge will take much longer than the lifetimes of most current agents, our commitment must be such that we act in consistency with the situation as if we were going to be alive when the challenge is eventually met. This commitment can be reflected in our practical self-understanding, if we identify with the narrative that arises from our experiences. Our narrative identity emerges through incorporating our actions, experiences, or psychological characteristics as our own into the self-told story of our life. For such narrative identity to have ontological relevance, we must consider ourselves as being numerically identical with ourselves at different times. In other words, although during my adult life I have had many different experiences that have formed my identity through two or three decades, I still identify with who I was in my early twenties. To be a genuine moral agent, an agent's experiences must be actively unified, must be gathered together into the life of one narrative subject. In this way, we can make sense of the practical concerns in the deliberative phase, since I can only anticipate how future experiences will affect me if they are mine (Shoemaker 2015, §2.3). Committing one's self to the long-term project of meeting Humanity's Challenge thus presupposes that one's agency extends over time.

Of course, there is a lot more to say about conceptions of agency, but doing so is beyond the scope of this book. For now, I have given a minimal account of how we must practically understand ourselves if we want to engage in sustainable action over a long period of time. This initial proposal for understanding narrative agency is compatible with how we understand ourselves as being an agent in social institutions. However, it does not allow for the delegation of difficult tasks to a future self, since the agent identifies with this future self just as she does with herself now. Although I might come across new arguments in the future that are in favour of or against taking sustainable action, I, then, will still identify with who I am today.

Let us now discuss other moral implications of the Reductionist View. Parfit asks whether changing our view about personal identity to the Reductionist View may have consequences for our moral convictions (Parfit 1987, 321). In response, Korsgaard concludes that Parfit's conception of the person is not morally neutral (Korsgaard 1996b, 386). Interestingly, Korsgaard reacts to the moral implications of the Reductionist View with regard to utilitarian attitudes about paternalism (§§107–108) and distributive justice (§111) (Korsgaard 1996b, 380). She, however, does *not* comment on the sections in *Reasons and Persons* in between these topics, which are devoted to desert (§109) and commitment (§110). In addition to Korsgaard's argument for the unity of agency, I believe that the Reductionist View has implications for legal and ethical concepts, such as desert and commitment, which we cannot accept. Furthermore, I would also like to extend the argument to responsibility and rights. If we want to prevent agents from delegating their decision to future selves, then the conception of agency we need for working on Humanity's Challenge must be in line with the one I have proposed above. The moral consequences of the Reductionist View give us additional reasons to be cautious about adopting it.

Let me demonstrate these various implications as follows. First, on the Reductionist View, desert for past actions varies with the degree to which one is psychologically connected to those actions. Although Parfit claims that it is possible on the Reductionist View to argue that persons should be punished for crimes they committed in the past, he finds it necessary to refer to compatibilist and incompatibilist views from the debate on free will, stating: "On the Compatibilist View, the kind of free will that is required for desert would not be undermined by the truth of Determinism. On the Incompatibilist View, Determinism undermines both free will and desert" (Parfit 1987, 325). On a Non-Reductionist View, one believes that personal identity involves an entity distinct from physical and psychological continuity. Parfit concedes that on this view the argument is defensible that only this entity carries with it desert for past crimes, and that, if there is no such entity, there is no desert. This is the analogue of the incompatibilist view on free will, and Parfit cannot find a decisive argument for desert based on psychological continuity on the Reductionist View only (Parfit 1987, 325).

Moreover, Parfit states: "When some convict is now less closely connected to himself [than] at the time of his crime, he deserves less punishment" (Parfit 1987, 326). This implies that the degree to which we can be held responsible for our actions depends on the degree to which we have been able to establish the various relations between the experiences in our life that led to those actions (Shoemaker 1996, 320). We ought not to welcome this implication of the Reductionist View. It would incentivise criminals to simulate only very loose psychological connections to their crimes, which would undermine the foundations of legal systems in the affluent countries. Accordingly, anyone considering taking on Parfit's view should be clear on the counterintuitive claims this commits one to.

Second, on the Reductionist View one cannot make sense of long-term commitments or obligations. As with desert for punishment, Parfit cannot decisively argue that we are bound by commitments made in the past, because a different self than our current self could have made the commitment. However, as Parfit argues, in the case of commitments and obligations the situation becomes more complicated, as the question of personal identity enters twice. We must consider both the maker of a particular promise, as well as the person to whom the promise is made. The maker of the promise might say: "I, and all of my later selves, shall help you." But a forceful objection is that I can only bind my present self (Parfit 1987, 327). It could, then, be argued that the receiver of the promise, too, will have successive selves. In other words, when two people enter into an agreement, twenty years later it will be unclear whether both persons are still bound to that agreement, since both could have changed to another self. In conclusion, the bookkeeping of promises and obligations, in relation to the successive selves people can have under the Reductionist View, becomes too messy for the concepts of obligation and promise to remain meaningful.

Third, on the Reductionist View it becomes unclear how responsibility can be defined and how we can explain firmness of character and personal integrity over a longer period. Responsibility is the response to a challenge that one has to face, and to which one has to respond in a rational and reasonable manner (Werner 2002, 521). We describe as responsible someone who is prepared to and capable of exposing herself to and responding to the criticism of others regarding her actions, where an omission also counts as an action. Responsibility presupposes a consciously performed action based on reasons, and, subsequently, the agent's acceptance of the consequences. The consequences of the action belong to the action itself. That is, the attribution of responsibility is only possible insofar as free decisions of autonomous agents can have a causal effect in the external world (Werner 2002, 524–525). How can this be made compatible with the Reductionist View, which has abolished the subject of experiences? As with the case of desert, I conclude that retrospective responsibility runs into the same sort of difficulty. The difficulty for prospective responsibility follows from the problem with commitment and obligation.

Fourth, the Reductionist View is not suited as a foundation for an account of rights. Recall that rights appear in the following form: A has, with respect to B, a right to X on the basis of Y (see §1.3). Hence, the question is: Can the concept of personal identity, which we must presuppose when we conceptualise the right's subject A and its addressee B, be made compatible with the Reductionist View? Within the context of the rights of future generations the argument runs as follows.

In order for B to accept restrictions to her agency in view of the moral obligation she has to respect the right that A will have in the future, B must presuppose that A is (potentially) able to exercise her right as norm Y prescribes. This means that B must attribute to A the (potential) ability to subjectively judge whether her right has been violated. We should thus conceptualise A to be (potentially) in the position to determine a discrepancy between her situation at a particular moment $t_2$ and how that situation could have been, had B acted differently at $t_1$. Consequently, we must attribute to A the (potential) ability to judge whether the discrepancy detected in the previous step implies a violation of her right to live her life as a human being above some minimal standard of wealth. And this presupposes that A (potentially) has a conception of minimal standards on how one's life as a human being ought to be, and whether her own life meets these standards. In addition, we must attribute to A the (potential) ability to extend the awareness of her own situation to her entire generation of human beings living at $t_2$. In other words, if we attribute rights to A and derive moral obligations from those rights for B, then we implicitly presuppose (potential) abilities in A and B, which in turn are based on capabilities of rationality and autonomy in human beings.

When returning to the question above, we should ask whether these abilities are still compatible with the Reductionist View. It seems inconceivable to think that A's and B's reasoning could take place without presupposing a subject of experiences, the same subject that has been eliminated from the Reductionist View. The difficulty arises due to the changing of selves and the volatility in connectedness to experiences that can happen under the Reductionist View. This would result in contingent conceptions of the minimal standards on how one's life as a human being ought to be and contingently changing interpretations by A of her own life against the background of that conception. In other words, at time $t_2$, A might claim that her right to a life as a human being is violated, whereas at time $t_2 + 1$ she might withdraw this claim because both her self and her conception of the minimal standards of a life as a human being may have changed in the meantime. The concept of rights needs a more stable concept of personal identity as a foundation than the Reductionist View can offer.

In summary, I have argued that the Reductionist View faces serious problems when confronted with ethical concepts such as desert, obligation, responsibility, and rights. It could, however, be objected that it does not matter if we consider current people according to the normative-phenomenological conception of agency and future people according to the Reductionist View. I will now argue that it would be incoherent and morally hypocritical to continue with two different conceptions of human nature: one for current people and a fundamentally different conception for future people.

When the rights of current people are questioned, in affluent countries we can rely on a morality and a legal system, including appeal procedures and independent courts, which are all based on respecting one's personal autonomy. This implies that we, as current people, consider ourselves as actual subjects of rights, for which we have designed and implemented a fine-grained legal and moral framework to deal with those people who violate our rights. If we conceptualise future people according to the Reductionist View, then we seem inclined to treat future people differently from a moral point of view than we do current people. This is so because the Reductionist View faces problems with ethical concepts such as desert, obligation, responsibility, and rights. It implies that we treat current people, as actual human beings, and future people, as potential human beings, differently when their rights are concerned. We seem to shift to another moral universe when it comes to violating the rights of future people, since violating rights under the Reductionist View is different from violating rights under the normative-phenomenological view.

Future generations or their ombudsman could accuse us of moral hypocrisy (see §2.3). We want to appear moral by wanting to work on Humanity's Challenge, but as soon as the cost of sustainable action becomes too high, we start talking about future people under the Reductionist View. This view considers the identity of future people as implicated by the actions of current people. As a result, current people are inclined to believe as follows: Since our actions constitute their identity, they will not regret what we have done. Hence, I can delegate my decision regarding sustainable action to a future self and continue with my life as before. Future people don't care about whether

I act sustainably now. In other words, the agent's motivation to decide in favour of the option for sustainable action is hampered.

Coming back to the question this section started with (i.e. What can be said about the motivating force of the obligations to respect the rights of future generations?) I conclude that as long as the Reductionist View is in circulation, it can serve as a ready-to-hand excuse for letting the motive to live up to our obligations to future people be overridden by the self-interested motive not to change one's behaviour. The Reductionist View facilitates an underestimation of what our commitment to meet Humanity's Challenge entails. To keep alive the prospect of individuals being motivated to pursue Humanity's Challenge, it is important to emphasise how controversial, speculative, and counterintuitive the Reductionist View is. That is, if we want to make the motive that is implied by the way we understand ourselves as agents compatible with the motive to meet Humanity's Challenge and prevent ourselves from delegating our decision to a future self, then we ought to understand ourselves along the lines of the normative-phenomenological view.[8]

## 4.5 Overview and diagnosis

This chapter started with a methodological refinement of the practical syllogism in relation to the Rubicon model. Practical deliberation about taking sustainable action means that the agent makes a decision about the minor premise of the practical syllogism that follows from the motive to act on Humanity's Challenge. I argued that the outcome of the deliberative phase is an all-things-considered judgement that forms the start of the commitment phase, in which the agent decides to make a commitment (i.e. the all-out judgement) to execute the proposed action. Thereafter, I provided examples from the empirical literature of ways of deliberation that show which obstacles agents need to conquer. There seem to be many ways in which agents could fail in being instrumentally rational. That is, many cognitive obstacles could potentially hamper agents in reaching an all-things-considered judgement in favour of a sustainable action.

Two major philosophical obstacles were addressed in the remainder of the chapter. Agents tend to take the point of view of common-sense morality as their default moral position. The potential impact of this in the deliberative phase is that the difficulty an agent has with vividly imagining future people as potential victims of the agent's current non-sustainable behaviour constitutes an intractable problem for deciding in favour of taking sustainable action. Agents seem to have great difficulty with feeling the motivational force of abstract concerns about future generations. This would seem to be problematic when it comes to our motivation to work on Humanity's Challenge, since this motive (see §1.4) relies on the agent's awareness of the challenge's transcendence of spatial, intertemporal, and theoretical boundaries. By investigating the agent's perspective in the relatively simple example of a wasteful drive in a sports utility vehicle, I argued for the difficulties of replacing such an experience with a different experience such as a bicycle ride. In order to convince agents in affluent countries to reduce their luxury emissions it seems that institutional help is needed. Institutional guidance is needed to show how agents can replace the experience of a wasteful drive by the experience of taking a sustainable action. New social norms must be promoted that steer agents towards acting sustainably.

The fourth philosophical question from the list in §1.3, namely the question about conceptions of the human being that we ought to adopt if we want to take seriously our efforts to meet Humanity's Challenge, was then dealt with. If the agent adheres to the Reductionist View on personal identity, she potentially runs into a motivational obstacle, if she concludes as follows: The identity of future people is formed through entirely contingent events; if they are born, they will owe their very existence to the actions that we will have taken. Hence, I can delegate my decision to a future self, and it does not matter whether I act sustainably now. In response, I argued that the Reductionist View has implications for legal and ethical concepts, such as desert and commitment, which we cannot accept. In particular, I extended the argument to responsibility and rights. The conception of agency we need for working on Humanity's Challenge must be in line with the normative-phenomenological conception of human agency that I have proposed. This is the view that agents are capable of integrating a sense of ownership of their actions with their aims, ideals, and aspirations. Agents deliberate, plan, and execute actions for which they take responsibility. Our narrative identity emerges through incorporating our actions and experiences as ours into the self-told story of our life. If we want to commit ourselves to take up Humanity's Challenge, we can resist temptations to delegate taking sustainable action to a future self, provided that we conceptualise the human being in line with the normative-phenomenological view.

## Notes

1 This is a well-known metaphysical conception that is applied to future people, e.g. when formulating the non-identity problem. The non-identity problem tells us that there is no moral difference between, e.g. conserving and depleting natural resources, because the identity of *future* people is fully contingent upon the actions undertaken by *current* people. See Parfit (1987).

2 Since we are considering voluntary and purposeful action, such action is always done for a reason. A basic distinction is between reasons that justify the action (normative reasons) and reasons that explain why the agent acted as she did (motivating reasons). A reason that motivated the agent may be unjustified from a normative point of view, and a normative reason may not be able to motivate. However, I follow Jonathan Dancy, who argues that normative and motivating reasons are not fundamentally different, but are meant to answer, respectively, two different questions: Was it a sensible thing to do? And, why did the agent act as she did? (Dancy 2000, 2). Normative reasons can be among the reasons that motivate, and motivating reasons can be the reasons that justify an action (Dancy 2000, 137). See also Scarano (2002, 432).

3 I base this presentation on Scarano (2002) and Gottlieb (2006). Moreover, I follow Davidson in that the conclusion of a practical syllogism can be an action, but it also "may be an intention to do something in the future" (2001, 86).
4 Moreover, to emphasise the practicality of practical deliberation, Gottlieb concludes as follows: "Finally, the minor premise of Aristotle's practical syllogism shows how and why one cannot have practical wisdom without ethical virtue and vice versa" (2006, 230).
5 Notice that Bratman considers Davidson's account of intention to be "a special kind of evaluation" (Bratman 1999, 110). In Chapter 5, we will use Bratman's theory of intention, which ascribes normative aspects to commitments that follow from an agent's future-directed intentions. This is of no concern, as considering Davidson's account of judgements here serves only to clarify the difference between the deliberative and the commitment phase. In Chapter 5, I will argue what commitment entails in addition to the all-out judgement.
6 Note that the authors define means as "actions, intentions, choices, decisions, etc. that might make use of such tools and resources. It is natural to think of means to an end, so understood, as actions that help to *cause* the end to occur" (2013, §2, emphasis Kolodny and Brunero).
7 The homo-economicus model in mainstream economic theory has been criticised by authors such as Sen (1977), Spash (2005), Becker (2006), and Ulrich (2008).
8 Elsewhere, I have argued that there are, in addition, metaphysical problems that follow from the Reductionist View (Mees 2010). In particular, Parfit's Time Dependence Claim, which has the form "If P, then he would in fact never have existed" (Parfit 1987, 352), begs a specific concept of personal identity, which has consequences for the way we conceptualise future generations and their right to a life as a human being.

# 5 Motivational challenges in the commitment phase

## 5.1 Introduction

In Chapter 4 we considered the agent in the mode of deliberation. We now follow her while she is committing herself to carry out the action that came out of the deliberative phase as the chosen one. This chapter looks at the agent while she is dedicating herself to that action, forming relevant intentions, and planning their execution.

We thus assume that, as far as the agent can be considered rational, she has reached the all-things-considered judgement about taking a sustainable action (see §4.2). The transformation that takes place in this phase is characterised as a resolution, resulting in felt determination to achieve the relevant outcome. It becomes an end-state that the agent feels committed to achieve, which constitutes a psychological resource for meeting the unanticipated challenges of the executive phase. Hence, unless there is new information that requires the agent to reconsider her choice of the action proposal (and thus to return to the deliberative phase), we may safely assume the agent to remain in the commitment phase of the Rubicon model.

We will consider two lines of philosophical theory in analysing the commitment phase: intrasubjective and intersubjective theories. First, making a commitment involves an intention to act. The forming of intentions is among the first signs of the agent's commitment. The concept of intention we will use comes from Michael Bratman's intrasubjective theory of intention and planning (Bratman 1999). Second, making a commitment to act morally leads to changes in one's social relations to other agents. We will consider the theory of recognition as developed by Axel Honneth as the basis for analysing this intersubjective phenomenon (Honneth 1995).

Thus, the structure of this chapter will be as follows. We will start by exploring the vast terrain of the phenomenon of commitment. First, I will describe the process of making commitments as a form of autonomous agency. Making a commitment means that the agent binds her will to take action. I will then argue that deciding to take action means forming an intention to take action in the future, which could be immediately. Following Bratman, I describe intentions as conduct-controlling pro-attitudes that display stability or inertia and are input to further practical reasoning. Now, in order to apply the theories of intention (and, later on, recognition) to the case material, we will need two measures for comparing them: to that effect, therefore, I briefly introduce the concepts of *requirement* and *risk*. With these ancillary concepts we will be able to discriminate between the cases (§5.2).

The two subsequent sections deal with specific features of intention, which involves looking into the case material to find differences between the sustainable actions and the other actions. In §5.3 we consider the requirements for internal consistency and means-end coherence as aspects of actions where potential differences can be found. In §5.4 we look at the belief an agent has in her own following through with executing the action she intends and consider the risk of intention-belief incompleteness. The structure of these two sections is similar: first some theoretical considerations are developed, and these are then applied to the case material.

The theory of recognition, as formulated by Axel Honneth, approaches human individuality on the basis of its being constituted through interaction with other human beings. The theory of recognition reminds us that, through being called upon in our relations to other autonomous human beings, we develop agentic resources that help us in keeping to our commitment. This theory has been specified further in an account of the agential competencies of self-trust, self-respect, and self-esteem (Anderson and Honneth 2005). As with the theory of intention, I will first give some theoretical considerations, and will then apply them to the case material (§5.5).

In §5.6, I give an overview of the findings of the previous sections with the aim of formulating a descriptive diagnosis of the commitment phase. The diagnosis of this action phase provides us with a tentative hypothesis, in answer to the question "Why is it hard to engage in sustainable action?", which is focused on making a commitment and the subsequent preparation of the executive phase.

## 5.2 Exploring the commitment phase

For the purpose of this chapter we assume that during the deliberative phase those reasons have been addressed that could have prevented the agent from deciding in favour of taking the action that is now waiting to be executed. The process of weighing reasons in favour and against action proposals has taken place, and the current option for action has come out of the selection phase all things considered. With the agent now free from possible deliberative hindrances, in this phase and the executive phase we can consider her problem with motivation in its purest form. That is, the agent has become aware that the proposed action will happen if and only if *she actually initiates* the chain of events pictured as the most desirable one in the deliberative phase.

The consequences of this awareness will be described in this section in two steps. First, I will argue that agents in the commitment phase face the problem of binding their will when they decide about action proposals. At the same time, I continue to focus on taking a decision for action as leading to intending future action. Second, I will discuss those elements of the theory of intention that are needed to start looking into how intention works in the case material. Before we leave this section, two ancillary concepts, requirement and risk, will be introduced. We need these to enable a comparison of the analyses of the cases in §5.3.

Given that it is now up to the agent to set things into motion, we shall discuss the question: What does it mean for an agent to make a commitment in favour of an option for action? That is, what happens when agents commit themselves to executing a proposal for action that has survived the phase of critical examination? This means that we now consider the agent, standing in front of the action proposal, facing just one question: Do *I* want to take this action? More specifically, do *I* want to move myself so as to become the cause of the projected chain of events that will result from my decision?

In the moment of coming to a resolution the agent realises that she is alone; she is thrown back on herself. There is no one else to rely upon in the moment of appropriating the proposed action to her. The agent faces the following choice: How will engaging in this action change *my* life? Knowing the simple fact that I alone can initiate my actions (Buss 2014, §1), am *I* prepared to do what is required of me in order to act efficaciously?

We have assumed that an agent's awareness of the experience of acting, while she is living through or performing it, makes it a conscious experience (see §2.4). This implies that when a morally substantive action is concerned, the agent realises that she is not a cog in a machine, but that the act will make her responsible for the moral consequences involved. She faces the moment of her resolve, assuming moral responsibility for what she is going to do. While making a commitment to act in accordance with Humanity's Challenge, it seems that an agent's moral responsibility is being called upon in a quite specific way: Am *I* prepared to do or refrain from doing certain things in order to meet the challenge, even though my actions will primarily benefit people who will live in the distant future?

Making a commitment, understood in the above sense, could mean many different things to agents. For example, becoming aware that it is up to her whether the action proposal will be realised could trigger two types of emotions. As they experience the limitations of their existence in space and time, agents realise that the consequence of their resolution in favour of a particular action means that other options cannot be carried out. The experience of choosing in favour of one action proposal with the result that other ends perhaps cannot be achieved could arouse feelings of regret and loss that the agent would like to avoid. On the other hand, the agent could recognise the proposed action as instrumental to realising her ideal of a good life. She could view the action as connected to what she believes to be worth doing, resulting in a strong feeling of ownership and up-to-me-ness that furthers determination and commitment.

Another potential obstacle that has bearing on the agent's decision may be her unreflective habits when making decisions. Although the agent has been rational during the deliberative phase, her way of appropriating the action to her could be governed by conventions the agent nonreflectively adheres to. Some of the nonreflected background assumptions that were scrutinised during the deliberative phase may pop up again, leading the agent to decide in favour of an option for action without being conscious of the sort of moral responsibility that is involved. For example, in choosing between options for commuting to work one could be influenced by the habit of taking the car instead of using a sustainable means of transportation. Although the upshot of the deliberative phase has been that sustainable commuting is the preferred act all things considered, the agent's habit causes her to make the all-out judgement in favour of taking the car (see §4.2).

As the above considerations make clear, whenever we act, there is no guarantee that, as Sarah Buss states: "the forces that move us owe *their* power to *our* power to decide what to do" (Buss 2014, §1, emphasis Buss). The number of possible influences on agents while they take decisions is simply vast. As I see it, when making a commitment, the agent faces the existential moment of binding her will to follow practical reason and to start executing the proposed action. Under a broad concept of autonomy understood as "self-rule or self-government" (Mele 2002, 529), when making a commitment, the agent faces the governance of her will. For our investigation, we want to understand the differences between sustainable actions and other moral actions as far as forming a commitment, seen as a binding of the will, is concerned. The theories of intention and recognition, to be relied upon in this chapter, serve as a means to get to a more detailed understanding of this existential moment for the agent.

Note, however, that after making her commitment the agent remains in a deliberative mode during the preparation of the executive phase. As the empirical research underlying the Rubicon model has shown, commitment has irreversible traits. The metaphor "the die is cast" has been chosen to indicate that reversing one's commitment may involve substantial effort. It seems that there is a cost to going back on one's commitment. This means that deliberation after having made a commitment is different from before. When deliberating after she has made her commitment, the agent takes on an attitude of nonreconsideration. Her commitment comes in when she binds herself to the action to a certain degree, despite having certain uncertainties or despite some aspects of the plan not being completely filled in. This, therefore, means that deliberation, after the commitment has been made, seems oriented towards finding solutions for the practical problems the agent may anticipate. Making a commitment seems to create a psychological resource, which enables the agent to overcome motivational roadblocks that could prevent her from continuing with the action.

Our basic question is: Why does taking sustainable action seem so hard from the perspective of the agent? From the practical syllogism (see §4.2) we take it that there must be coherence between the agent's motives for deciding in favour of the action proposal (minor premise) and the normative reason to help meeting Humanity's Challenge (major premise). As we will see below, in Bratman's theory of intention this coherence implies that the motive for taking the action must fit with the agent's more or less long-term plans (Buss 2014, §2). If we did not assume this coherence, then we would not be able to say anything about how it is for the agent to take sustainable action, since the goal of a sustainable future and the agent's motive for acting would then only relate to each other in a contingent way. Hence, when the agent in the deliberative phase has selected an end E (e.g. acting on Humanity's Challenge), and if act X is constitutive of doing E, then the agent commits herself in the commitment phase to do X. The coherence between the agent's motives and her attitude towards the proposed action thus tells us that the agent's resolution results in an intention to do X in the near future. Deciding in favour of act X thus means that the agent has formed a future-directed intention to do X.

Our considerations so far have implications for which theory of intention would be suited for investigating the differences between sustainable and other morally relevant actions. For example, we could consider the reduction of one's future-directed intention to fit within a desire-belief theory. This means that instead of my intention to place solar panels on the roof of my house, I have the desire to do so. Such reduction, however, does not seem compatible with commitment, as I understand it, for two reasons. First, my desire does not direct my deliberation in the same way as my intention does. My intention directs deliberation towards settling the issue of placing the panels, whereas my desire allows me to keep deliberating as to whether I will install the panels. Second, as we will see below, intentions are conduct-controlling pro-attitudes, whereas desires are mere influencers of action. My intention, followed by the plan, to install solar panels directs my conduct towards looking for appropriate offers from suppliers, whereas my desire does not have this conduct-controlling effect; other desires could pull me in other directions (Bratman 1999, 18–19). We will now take a closer look at Bratman's theory of intention and planning, as this theory seems suited to interpreting the agential phenomena in the case material.

Ever since the early days of this theory (notably Anscombe 1957 and Davidson 1963), intention has been understood as appearing in three forms: intention for the future, as when I intend to finalise this book before a certain date; the intention with which someone acts, as when I am typing with the further intention to write this sentence about the three forms of intention; and intentional action, as in the fact that I am typing these words intentionally. The principal task of a theory of intention is to unify these three forms (Setiya 2014, §1). Here, however, we are interested specifically in characteristics of future-directed intentions, and not in an all-encompassing account of intention. I thus do not need to spell out a full theory of intention in order to find the characteristics of sustainable actions relative to other morally relevant actions. As one can see in the case material set out in Chapter 3, these action situations are complex. The actions extend over time, and they require the agent's persistence to follow through with them. To understand how intending future action works for the agents, acting in these cases, I highlight three elements of the theory of intention, which Michael Bratman has presented in detail.

First, according to Bratman, intentions are conduct-controlling pro-attitudes that exhibit inertia and function as inputs for further practical reasoning. This is a technical definition for which Bratman provides the following content (Bratman 1999, 20). A future-directed intention is understood as a commitment to take action in the future. A characteristic of commitment that we need to consider concerns the relation between intention and action, which Bratman calls "the *volitional* dimension of commitment" (Bratman 1999, 15, emphasis Bratman). Intentions and desires have in common that they are pro-attitudes. Pro-attitudes play a motivational role – that is, in combination with our beliefs they can move us to act. Now, Bratman adds that, in contrast with ordinary desires, intentions are *conduct-controlling* pro-attitudes, whereas desires should be regarded as mere potential influencers of action. If an agent's future-directed intention manages to survive until the time of action, and the agent notices that the time has arrived and nothing interferes, then it will control the agent's action (Bratman 1999, 16). This means that when I have the intention to reduce the carbon emission of my household by switching in part to solar energy, then I will accept a suitable offer from a supplier of solar panels. If I were only to desire to reduce the carbon emission of my household, this in itself would not be sufficient for me to accept the offer of the solar panel supplier.

Second, intentions have stability or *inertia*. After forming a goal intention, people start preparing for action execution. The phenomenon associated with this activity is planning. Since agents operate in complex environments, planning is often necessary, because newly formed goal intentions cannot be implemented immediately, for example if the agent is engaged in alternative activities that first need to be completed or if relevant opportunities to act are not yet available. Intentions play a role in the period between their initial formation and their eventual execution.

If I intend to put solar panels on the roof of my house, under normal circumstances I will not continue deliberating about whether I will do this. My intention to switch to solar energy resists reconsideration, under normal circumstances. When I become aware of new information, however, for example that the producer of my solar panels has gone bankrupt and that I can buy from another supplier only at a much higher price, I may return to the previous action phase and reconsider my original intention. Assuming all conditions to be equal, however, the default situation is the retention of my prior intention – my "nonreconsideration" of it, as Bratman puts it (ibid.).

Third, intentions are inputs to further practical reasoning. When I intend to switch to solar energy for my house, I will further plan this change in that I will, for instance, look for ways to apply for a subsidy under a governmental sustainable energy scheme (if applicable) or organise for someone to be at home when the supplier arrives on the day of installation. In other words, my original intention gives me a variety of further intentions that form necessary means to executing the original intention. I will frequently reason from intended end to intended means or preliminary steps (Bratman 1999, 17). My further practical reasoning will also restrict the forming of other intentions I may consider, for example, spending my next holidays on an expensive cruise. Realising my desire to go to on this cruise would cost me the money I have decided to spend on the solar panels. My intention to install the solar panels may thus prevent me from intending to go on an expensive cruise for a holiday.

So far, we have established a view on what it means to commit to an action proposal, and we have related this view to the theory of intention, of which we have highlighted a few important aspects. When we now turn to review the case studies with regard to the specific aspects of the intentions involved, we will need to assign certain qualifications to these aspects, such that they can be compared across the cases in a more or less univocal way. Therefore, I will introduce two ancillary concepts that enable us to compare the cases with each other as regards the relevant aspects of intention.

First, the notion of a *requirement* will be used. A requirement, in the sense that I use it here, is a pragmatic necessity that comes in degrees. An aspect of an intention is scored according to the extent to which it is required. This means that the necessity of this requirement is not absolute in a logical sense (i.e. binary); but it is meant to vary in degree. A requirement can be a *must have*, in the sense that it must be satisfied for the action to be considered successful. Such a requirement is absolute; without the requirement being fulfilled, the action cannot be executed. A requirement can be a *should have* – when it has high-priority and should be included in the action if possible. Such a requirement is often critical for successfully carrying out the action, but is one which can be satisfied, if necessary, through alternative ways of acting. A *could have* requirement is one which is considered desirable but not necessary. Such a requirement will be fulfilled in carrying out the action if time and resources permit. In summary, the need for a certain aspect of intention can be expressed qualitatively as a must, should, or could have of that particular aspect.

The way in which the notion of requirement will be used is as follows. Suppose that we evaluate the cases on the aspect of means-end coherence in the agent's plans. We might then say, for example, that it is a 'must have' in case P and a 'could have' in case Q. For example, the doctor discussed earlier has committed himself to join Médecins Sans Frontières (MSF) (see §3.3). As MSF is an established institution, it has implemented a procedure for new doctors joining the organisation. The doctor is supposed to rely on the means-end coherence as prescribed by MSF, and not to find out for himself. I consider the requirement for means-end coherence for this doctor, therefore, a 'could have'. If we, then, compare cases P and Q regarding the aspect of means-end coherence in the agent's plans, this will lead us to diagnose that in case P agents need to work harder on the means-end coherence of their plans than agents in case Q. In this way, we obtain a first impression of a comparison of cases P and Q (in relative terms) on the aspect of means-end coherence.

Second, some aspects of intentions need to be described in terms of a *risk*. In this book, we are looking for a robust concept of risk that allows us to distinguish between various morally relevant actions. For a definition of this concept, I follow Aven and Renn, who evaluate ten different definitions of risk in order to finally propose: "Risk refers to uncertainty about and severity of the consequences (or outcomes) of an activity with respect to something that humans value" (Aven and Renn 2009, 6). For our purposes, risk refers to uncertainty about the consequences of an event, seen in relation to the severity of these consequences. *Severity* is a way of characterising the consequences. It refers to intensity, size, extension, scope, and other potential measures of magnitude – with respect to something that human beings value (lives, the environment, money, etc.). We express the uncertainty involved by assigning a *probability*, for example by estimating the probability of an event as high. Since I am not interested, here, in subtler quantifications of risk, I express risk as the combination of the severity and the probability of an event. Hence, as risk is defined as the two-dimensional combination of uncertainties and severity of consequences, any judgement about the level of risk needs to consider both dimensions simultaneously.

When assessing the relevant aspects of intention, I will assign broad labels to both severity and probability, by qualifying them as low, medium, or high. A risk is considered high when both severity and probability are high, or when one of them is high and the other is medium. It is considered medium when both severity and probability are medium, or when one of them is high and the other is low. It is considered low in all other cases.

## 5.3 The need for planning

We have already observed that the actions in the case material are temporally extended; they require the agent's persistence to follow through. Hence, at least in such cases, we can concur with Bratman that we are planning agents (Bratman 1999, 2). Having a plan that guides our later conduct seems crucial for agents involved in complex action situations.

We will need to understand the specific characteristics of those plans in the case of sustainable actions and in other morally complex situations. Plans and intentions relate to each other in the sense that future-directed intentions are part of larger plans. Characteristically, these plans play roles in the agent's coordination and ongoing practical reasoning (Bratman 1999, 8). Like intentions, plans are conduct-controlling, they resist reconsideration, and they provide crucial inputs for further practical reasoning and planning. But, as a result of their more complex nature compared with the simple structure of intentions, plans have two further typical features.

First, they are *partial* in the sense that when I decide to place solar panels on my house, I have some picture of what this entails, but, for example, I do not know in detail when the supplier will do the installation, or where exactly the electricity converter will be placed. Hence, although I am aware that my plan is incomplete, I can fill it in later as required. Second, plans have a *hierarchical structure*. While my plan to switch in part to solar energy remains as it is, I may start a new deliberation about the aesthetics of solar panels on my roof, resulting in the choice of mixed grey ones (say). That is, one may hold certain intended ends fixed, while deliberating about means and preliminary steps towards realising those ends (Bratman 1999, 29).

For our analysis of the case material it is important to know the demands on plans that need to be satisfied. First, to coordinate the agent's activities over time, a plan must be, other things being equal, *internally consistent*. In particular, it needs to be consistent with one's beliefs about the action: it must be possible for the agent's entire plan to be successfully executed assuming that her other beliefs concerning the action are true (Bratman 1999, 31). This demand implies that my plan to place solar panels should be consistent with my belief that putting solar panels on my house is physically possible. I cannot plan to put solar panels on my house while also believing that the roof construction is irredeemably inadequate: that would violate the demand for internal consistency, and thereby make it quite hard indeed to execute the plan. As outlined in §5.2, the requirement for the internal consistency of plans can vary in degree, and, therefore, can be described in terms of a 'must', 'should', or 'could have'.

Second, there is a demand for *means-end coherence*. My plans need to be supported by sub-plans concerning means, preliminary steps, and more specific courses of action. Prior to the installation date for the solar panels, I need to read and sign the contract, and I have to make arrangements so that the workmen can access the roof. These sub-plans must be instrumentally coherent with the overall plan to switch to solar energy (see §4.2). "Failure to fill in my plans as needed in these ways will leave them means-end *in*coherent" (ibid., emphasis Bratman). In the need for more detailed sub-plans, the experienced agent can be distinguished from the inexperienced. For implementing complex ends, experienced agents can do with relatively high-level plans, since they have already formed relevant habits during earlier experiences that allow them to recognise recurring patterns in a given action situation. As we can see in the case studies in Chapter 3, these agents all bring with them various degrees of experience from which the need for more- or less-detailed sub-plans follows. As with internal consistency, we can assess the requirement for means-end coherence in terms of a 'must', 'should', or 'could have'.

Since we are looking for a characterisation of motivation for sustainable action, I will now review the case material with regard to the demands of internal consistency and means-end coherence. Regarding internal consistency, we should note that this concerns a very basic form of consistency. The agent's plan must have a basic fit with her beliefs about the world. The action situations are moral in the sense that agents act in ways that can be judged good or bad in the eyes of other agents who are affected. On the demand for internal consistency all cases seem to be alike; it is a 'must have' that this consistency is in place. It seems hard to imagine that a morally relevant action of the sort we have in mind could be taken which is inconsistent with one's beliefs. Carrying out a plan with the intention to $\varphi$, believing that $\varphi$ requires $\psi$, whilst also believing $\sim\psi$, seems the more doubtful given that the examples of actions all require social coordination with other agents. Acting in social contexts without internal consistency of one's plans will most likely lead to a lack of credibility, if not to immediately becoming criticised for irrationality, resulting in inefficacious agency.

When considering the requirement for means-end coherence, the above picture of uniformity across the cases changes. In cases where there is a potential supportive reception of the agent's plan to undertake the action, the requirement for means-end coherence of the sub-plans can be established by socialising with other agents. The semi-vegetarian (see §3.2), for example, could be in the position to outsource her intention to reduce her meat consumption to a partner who happens to take care of the daily shopping and cooking (i.e. a 'should have').

As noted in the previous section, the doctor without borders has decided to join MSF (see §3.3) and as MSF is an established organisation, there is a procedure in place for new doctors who will be sent on the missions. The doctor is supposed to rely on the means-end coherence as provided and prescribed by MSF, and not to make up his own plans. Hence, the doctor without borders does not seem to be under much pressure to care about means-end coherence – it is outsourced, so to speak, to the procedures of the MSF (i.e. a 'could have').

The credit restructuring officer in a bank (see §3.6) is working in the interest of the stakeholders of her organisation. For each of the alternatives in her action situation (continuing or ceasing to lend to a customer), although they have a significant moral impact, detailed plans are largely dictated by legal constraints. This implies that the means-end coherence of her intention in action is in part controlled by the legal and organisational context in which she is operating. Should the detailed plans not be available, there would be ample support from colleagues and superiors to draft them and keep them means-end coherent. However, given the strong interests of opposing parties involved, I deem the need for means-end coherence a 'should have'.

Thus, in these three cases (semi-vegetarian, doctor without borders, credit restructuring officer), the agent does not seem to be required to establish means-end coherence *herself* each and every moment, since there are possibilities for relying on her social environment to do this.

The situation, however, is different for the "power grid rebels" and the purchase manager (see §3.4 and §3.5 respectively). From the case descriptions, we observe that these agents encounter resistance when they make public their plans on how to achieve the sustainable ends they want to reach. Despite turning the public opinion towards renewable energy, the "power grid rebels" face strong stakeholders in opposition. Likewise, the purchase manager faces scepticism from senior colleagues, who have the power to end her project. A common characteristic seems to be that these agents do not have an organisation that could take over the task of filling in their plan. Their project stands or falls with *them*, as the ones who need to settle means-end coherence all the time without having the opportunity to rely on other agents who share the same intention or form part of the same institution. In these two cases, means-end coherence is a 'must have' requirement for the agents.

To summarise, there seems to be a meaningful difference between the sustainable actions and the other actions as far as the requirement for means-end coherence of the agents' plans is concerned. The agents who have committed themselves to act sustainably need to establish this coherence first and foremost by themselves, whereas the other agents can rely on support from others. This means that already in this action phase the agent's commitment to help meeting Humanity's Challenge seems to be different from her commitment in the other action situations. In a similar way, we will now investigate how this looks with regard to the agent's belief that she will follow through.

## 5.4 The belief that one will follow through

Making a commitment means that the agent goes through the existential moment of binding her will to follow practical reason and to start executing the proposed action (see §5.2). She may, nevertheless, have doubts about whether she will succeed. The purpose of this section is to search for specific features of the sustainable actions as regards the agent's belief in her effectiveness. As intentions and plans are conduct-controlling pro-attitudes, have characteristic inertia, and play a crucial role as inputs into and constraints on further practical reasoning, an important question is to what extent we may expect an agent to execute an intended action. That is, does my intention to $\varphi$ actually require a belief that I will $\varphi$?

As with the means-end coherence of our plans (see §5.3), our effectiveness in following through on an intention is partly a function of our degree of commitment to the course of action, particularly in light of what we anticipate the difficulties to be. An agent's commitment forms a psychological resource, which is required to cope with the foreseeable motivational challenges during the executive phase. By making her commitment the agent prepares to resist anticipated temptations that may keep her from following through. The commitment phase is the phase in which we resolve ourselves to persevere in the face of anticipated difficulties. It involves the formation of an attitude; an attitude formed in the light of our sense of the risks that our plan will be inadequate, since contingencies will thwart our efforts. Below, in order to approximate the agent's attitude of perseverance, I will make use of

Bratman's concept of intention-belief incompleteness to assess the case studies for features of the agent's belief that she will follow through.

Note that Bratman's theory of intention does not entail an affirmative answer to the question whether my intention to φ actually requires a belief that I will φ. And there is a reason why such a connection between intention and belief should be rejected as overly strong. Bratman illustrates this by referring to cases of absent-mindedness and agnosticism (Bratman 1999, 37). As an example of the former case, I intend to buy bread for my neighbour when doing my own weekly shopping. Since I am aware of my tendency towards absent-mindedness, my belief that I will follow up on my intention could be weak. It is not that I believe I will not buy bread; I just do not believe I will. The agnosticism examples concern situations where I am agnostic about whether I will succeed when I am trying to realise my intention. This happens, for example, when I face situations for which I am ill-equipped. When acting on my intention involves leading a large-scale war against criminals, I would do my utmost at each and every step. But since I am aware of my lack of knowledge and experience regarding fighting criminals, I do not believe that I will succeed.

To take these examples into account, Bratman develops his theory of intentions and plans in such a way that he does not require the strong assumption that to intend to φ I must believe that I will φ. Instead, he develops the following formulation:

> I will suppose that a normal role of an intention to φ is to support an expectation that one will φ; and I will also suppose that there is, other things equal, an important kind of irrationality involved in intending to act in ways inconsistent with one's beliefs.
>
> (Bratman 1999, 38)

Bratman defines the case of intending to φ without believing one will φ as *intention-belief incompleteness*. From this definition, one can argue that intention-belief inconsistency is closer to criticisable irrationality than is intention-belief incompleteness. The agent can be considered irrational when she intends to φ while believing that she will not φ. On the other hand, it does not need to be irrational for the agent to intend to φ while having serious doubts about whether she will φ.

For the morally complicated actions we consider in this book, we need to ask whether there can be a failure of the agent to execute her plan due to a situation of intention-belief incompleteness. When executing complicated plans, the agent will be faced frequently with highly contingent situations that could give rise to feelings of uncertainty about the efficacy of her agency. It is not that the agent believes she will not carry out the action; she just does not believe she will be successful for some reason that may emerge from the contingent situation she is in. We can, therefore, argue that there is a risk that the action execution will stop at some moment due to the agent's intention-belief incompleteness being such that she gives up.

Among the cases there seem to be some remarkable differences on the aspect of intention-belief incompleteness. As an institutional agent, the credit restructuring officer (see §3.6), although faced with clear moral dilemmas, is unlikely to be hampered by intention-belief incompleteness. Assuming that she has been selected for this job on the basis of her competencies, she knows the alternative scenarios in her action situation, and the probability of intention-belief incompleteness will be low. The organisation supports the credit restructuring officer in coming up with a solution to her dilemma; hence the severity seems limited: if the agent suffers from intention-belief incompleteness, then someone else's plan will replace her plan. Hence, the risk for intention-belief incompleteness is low.

The sustainable action taken by agents in an institutional role, however, shows a different pattern from the agent's point of view. Clearly, regarding support from within her professional environment for her plan, the purchase manager (see §3.5) is not in an enviable position. The complexity of the organisational change that is required, combined with the resistance she encounters are strong reasons to consider the probability high that she will simply not believe she will be able to follow through. As her efficacy in this matter entirely depends on the robustness of her plan, including the organisational tactics she may need to perform, the severity of her intention-belief incompleteness is also high. Thus, the risk for intention-belief incompleteness is high.

The situation regarding intention-belief incompleteness for agents in individual roles seems less clear-cut than for agents in institutional roles. Although the semi-vegetarian (see §3.2) can organise support and assistance for her plan, she faces the other side of her commitment to change towards a different kind of behaviour. This is that she has to live up to it, which under certain circumstances (e.g. carnivore friends), or given the agent's own psychology, may turn out to be difficult. Assuming that the situation is such as to not make the case trivial in this respect (e.g. by her being the last non-vegetarian in her community), we may assess probability to be at least medium, and similarly for severity, since no one else is there to achieve her end for her. Therefore, the risk of intention-belief incompleteness is likely to be medium (if not high).

For the doctor without borders (see §3.3) this risk is probably high. The case description tells us that circumstances are extreme. There is, for example, a high level of burnout among MSF staff. Knowing all this, the doctor will assess the probability of intention-belief incompleteness as high. However, since MSF is aware of this, it has a practice of relocating staff in cases of poor performance. Therefore, the severity may be judged medium.

Finally, the "power grid rebels" (see §3.4) face an immense task, which becomes even more complex along the way of execution. They need to gain control over the power grid in their town, whilst facing hostile opposition from an electricity company that has greater financial resources at its disposal. In addition, this fight takes place against the background of a lack of clarity about the support the agents will obtain from the other citizens in the town. The probability of intention-belief incompleteness is thus high. Severity is also high, since giving in to the belief that they will not succeed will almost immediately lead to inefficaciousness on the side of the agents. Their project starts out with a lack of support, and if the agents do not believe that they will succeed, nobody will step in to realise their end. For these reasons, I consider the risk of intention-belief incompleteness high.

In this section, we have considered the aspect of intention-belief incompleteness, which tells us the extent to which the agent feels confident that she will follow through. Since working on Humanity's Challenge does not seem to be a priority for many people in the affluent countries, it seems important to know whether sustainable actions are more demanding in this regard than other action situations. From the assessment above, our first diagnosis is that the agents taking sustainable action are more vulnerable than the other agents as far as intention-belief incompleteness is concerned.

## 5.5 Commitment and recognition

So far, we have concentrated on aspects of the agent's commitment in the sense of binding her will, which indicates that reversing her commitment may involve substantial effort. As stated in the previous section, an agent's commitment forms a psychological resource, which is of crucial importance for coping with anticipated difficulties in the executive phase. We will now concentrate on a theory that sees this agentic resource as being constituted intersubjectively through the social interaction with other agents. One's commitment can be affected by reactions from others to one's avowed intention to take moral action. A significant other who criticises your choice can easily challenge your commitment, especially if it is still in an early stage of formation. Criticism of my commitment by people whom I deem important may thus seriously affect this recognitionally secured psychological resource (Anderson 2014) on which I draw during the executive phase.

This section is about the agent's commitment and the reactions from significant others to the agent's expression of her commitment. As we have seen, in particular in the paradigmatic cases of the "power grid rebels" (see §3.4) and the purchase manager (see §3.5), these agents have to cope with resistance from other agents who do not support their plans. Commitment by one agent and resistance from other agents leads to the question how these agents relate to each other in terms of recognition. This is what we investigate here. We draw on Axel Honneth's theory of recognition without any pretension to pay tribute to the rich history of this theory. In that sense, we make use of just a limited part of the literature.

A first-person formulation of autonomy runs as follows: to be one's own person, to be directed by considerations, desires, conditions, and characteristics that are not simply imposed externally upon one but are part of what can somehow be considered one's authentic self (Christman 2015, §1). However, autonomy has important intersubjective characteristics as well. How do agents recognise each other if a commitment made by one of them generates conflict among these agents? How may such a conflict work out if these agents are working in the same social institution? In reply, we consider Honneth's theory of recognition, which argues that agents not only assert autonomy in liberal democracies, but that autonomy evolves as a result of the mutual recognition by other human individuals. In order to understand the patterns of recognition that are apparent in our assessment of the case material later on, we provide some of the elements of this account of intersubjective autonomy.

Recognition as an intersubjectively shared quality is understood as a threefold relationship: X recognises Y as Z, whereby X and Y could both refer to individuals and to groups, and Z forms the evaluative point of view under which the recognition takes place (Halbig 2002, 297). The definition implies that there can be consequences for both the moral and legal status of Y depending on the recognition by X, as well as consequences for her actions that X should accept following her recognition of Y.

For example, if X recognises Y as a professional colleague in an organisation, this means that X attributes certain competencies to Y that enforce Y's status as a professional agent. As a result, X takes the obligation seriously to respect Y as a professional colleague. The recognition by X of Y means that Y's perception of herself in the interaction with others like X is compatible with the perception agents like X have of Y as a professional colleague. Y perceives herself as a professional agent, and X also perceives Y as a professional agent. Likewise, when X ends her recognition of Y as a professional agent, this could lead to a different treatment by X of Y, and an incompatibility between Y's self-perception and X's evaluation of Y could emerge. This change in recognition could form the basis for a social conflict initiated by X, who starts acting against the commitment that Y has made.

Hence, the question I will address below is whether agents who undertake a sustainable action are in some way recognised differently than their colleagues in the organisation who do not take such action. The recognitional account of autonomy as formulated by Joel Anderson and Axel Honneth argues that

> the agentic competencies that comprise autonomy require that one be able to sustain certain attitudes towards oneself (in particular: self-trust, self-respect, and self-esteem) and that these affectively laden self conceptions … are dependent, in turn, on the sustaining attitudes of others.
>
> (Anderson and Honneth 2005, 131)

One's relationship to oneself, in particular one's competencies of self-trust, self-respect, and self-esteem, is thus influenced and even constituted by one's intersubjectivist relationships to other human beings. This means that one's efficacy in pursuing plans and projects relies heavily on one's recognition of other agents and on one's being recognised by other agents. Below, I will elaborate the conceptions of self-trust, self-respect, and self-esteem as psychological resources to the extent needed to apply them to the case material.

*Self-trust* "has to do with the affectively mediated capacities by which what is subjectively felt becomes material for deliberation in the first place" (Anderson and Honneth 2005, 133). Without self-trust we do not have access to our feelings, yearnings, fears, regrets, and so on. A lack of self-trust diminishes one's possibilities to live according to one's ideal of a good life, because one can no longer rely on unconsciously originated feelings as a source of one's most authentically held beliefs. Self-trust can be intersubjectively influenced positively or negatively: it can be impaired by rape and violence; it can be strengthened by intimate loving relationships, especially in early childhood. Self-trust, as the capacity to use and accept the unconscious resources of one's inner life, is formed in the early years of our lives in the way we are intersubjectively treated by parents, teachers, and others who influence this constituting part of our autonomy. If this capacity stays underdeveloped, or even handicapped, one will face difficulties in pursuing what one considers a worthwhile life.

*Self-respect* can be seen "as the affectively laden self-conception that underwrites a view of oneself as the legitimate source of reasons for acting" (Anderson and Honneth 2005, 132). It is a necessary condition for autonomy that we conceive of ourselves as capable of generating reasons for action, since without such a resource we would not be able to take ourselves seriously in our practical reasoning about what to do. Self-respect makes agents believe in their competencies as (professional) agents. Through self-respect agents become the co-creators of initiatives in social cooperation with other agents. As self-respect is to be viewed as being formed intersubjectively, it seems obvious that this psychological resource is vulnerable to subordination, marginalisation, and exclusion. Exercising the competency of self-respect depends largely on the respect of other agents. Our autonomy is hampered when we are faced with subtle or even radical changes in how other agents respect us, for example in our role as employee in an organisation. Although in constitutional democracies individuals are protected from certain basic forms of disrespect by a just legal framework, there remain numerous possibilities for changes in the relational patterns of respect between agents resulting in subordination, marginalisation, and exclusion of agents.

Someone who has sufficient self-trust and self-respect at her disposal can still be hampered in her autonomy if her acts (including speech acts) are misunderstood by her social environment. Indeed, acts of denigration and humiliation by other members in one's network of agency can be the reason for such a lack of understanding and consequent lack of opportunities to pursue one's ideal of a good life. In other words, in order to realise full autonomy one needs *self-esteem*; one needs "an especially high level of personal resilience, subcultural support, and persistent effort" (Anderson and Honneth 2005, 136). Without self-esteem – that is, when other significant agents do not support one's maxims – it becomes hard to pursue one's conception of a worthwhile life wholeheartedly. A socio-cultural environment that is hostile to what one considers meaningful and significant causes a deterioration of an agent's self-esteem and thereby her autonomy.

Having elaborated the concepts of self-trust, self-respect, and self-esteem, we can now investigate whether there is a meaningful difference between sustainable actions and other morally complicated actions as regards these psychological resources. Note that since the basis for self-trust seems to be formed during the early years of one's life, I take it that this psychological resource forms as much a necessary condition for taking a sustainable action as it does for any other morally relevant action. Moral action in general requires self-trust. Considering the competency of self-trust as being equally important for sustainable actions and other moral actions is consistent with Honneth, who argues that love, as the most basic form of recognition that contributes to an agent's competency of self-trust, "does not entail moral experiences that could lead, on their own accord, to the formation of social conflicts" (Honneth 1995, 162). Observing, therefore, that (a lack of) self-trust in moral actions will not differentiate between sustainable actions and other actions, I will concentrate on self-respect and self-esteem and review the case studies from Chapter 3 accordingly.

In addition to the conceptions introduced in §5.2, we need one more ancillary concept. The effect of a moral action on the agent's autonomy will be assessed in terms of her *vulnerability* to her autonomy being impaired as a result of diminishing self-respect or self-esteem. As vulnerability can vary in degrees, we can speak of an agent's autonomy having a low, medium, or high vulnerability to being impaired as a result of a hampered self-respect or self-esteem. Below, we first discuss self-respect, seen as a recognitionally secured psychological resource. This implies that, conversely, misrecognition by significant other agents leads to diminished self-respect, which may undermine one's commitment and resolve.

The commitment of the semi-vegetarian (see §3.2) to stop eating red meat seems to have most impact on her direct private partners, family, and friends. For these close members of the agent's network it may come as a surprise that she intends to change to another diet, but it could also be a common decision to change to low-meat. Supposing that her relationships to the network members are good, then, other things being equal, it seems unlikely that the agent will be excluded as a consequence only of her commitment to the new diet. Difficult situations may come up, due to the household members not sharing the same intention (e.g. two different meals having to be prepared instead of one). It will, however, be unlikely that the agent, as a result of her commitment, will no longer be respected in her role as partner, family member, or friend. The semi-vegetarian can continue in almost all of her roles (perhaps not as a cook for carnivore friends) without being subordinated or marginalised. Hence her vulnerability to an impaired self-respect is low.

The doctor without borders (see §3.3) works under harsh conditions, as a result of which he may run into psychological problems when he is on mission. However, this in itself does not make the agent a bad surgeon in the view of his colleagues. As the circumstances are extreme for everyone, that would not be a reason for him to suffer from an impaired self-respect. After a while the doctor or MSF may conclude that he may simply not be the right person to work with MSF. This, however, is unlikely to affect his self-respect as a surgeon. Thus, I assess the agent's vulnerability to a diminished self-respect as medium but not high.

Likewise, the "power grid rebels" (see §3.4) face difficult circumstances and resistance from powerful players. Their opponents may exclude them from certain local cultural activities. However, being protected by their rights in an affluent country, it is hard to imagine that they would not be able to continue in their various roles as a result of their commitment to realise renewable energy for their town. Hence their vulnerability to an impaired self-respect is medium at the most.

The credit restructuring officer (see §3.6) operates in a strongly embedded way in an organisation that supports her in her role. There is a high awareness that her job is difficult in the sense that it involves morally dilemmatic situations. This means that mistakes are part of the job. However, as long as the agent creates enough support for her actions, her self-respect remains intact. Even if she were to make a mistake that was not supported, there would be empathy among her managers, who would arrange a safe landing elsewhere in the organisation. This means that her motivation for engaging in her morally complicated actions is hardly affected by a potentially diminished self-respect. Thus, I see a low vulnerability to diminished self-respect in this case.

Concerning an agent's vulnerability to impaired self-respect *and* self-esteem, I consider the case of the purchase manager (see §3.5) to be paradigmatic. Here we observe an agent's narrow escape from actual exclusion from the company (in the form of a dismissal) as someone who is no longer considered as having the competencies of a professional purchase manager (self-respect). Simultaneously, her ideal of a worthwhile life, of which the change to sustainable palm oil is part and parcel, is criticised severely by people with a more senior position in the organisation (self-esteem). Only when the CEO of FrieslandCampina intervenes and puts the weight of his position behind the plan do things begin to change in a direction that restores the agent's resources of self-respect and self-esteem. As a result, the purchase manager sees her moral status in the company start to improve after a considerable period of feeling subordinated and denigrated. In my view, this case is the clearest example of what it means to undertake a sustainable action and have misrecognition by others at least temporarily affect one's self-respect and self-esteem. The vulnerability in this case study to both an impaired self-respect and an impaired self-esteem is high.

In the other cases, things seem to be slightly different when we consider the potential impairment of self-esteem. The semi-vegetarian can run into situations in which it is difficult for her to remain persistent as a result of denigration and humiliation by other agents. For example, this could happen if friends, who do not underwrite her policy, start to redefine the friendship by excluding the topic of sustainability from their commonly shared values. Then, the agent's self-esteem diminishes, wherefore her vulnerability to an impairment of self-esteem may be judged medium.

On the other hand, the doctor without borders will not be confronted with these attitudes. Assuming that he has taken his decision on the basis of support from his loved ones, the agent will receive substantial moral support from his entire network for his commitment to deliver emergency aid. Even if he were not able to actually perform the job of doctor in an emergency, his commitment to MSF would not be criticised in any way. The doctor would then need to find other ways of pursuing his ideal of a good life (e.g. by donating to MSF). However, his vulnerability to a diminished self-esteem, in my view, is low.

The example of the "power grid rebels" shows that, indeed, there can be a high vulnerability to being hampered in realising what one finds meaningful in life if one has to deal with the obstacles these agents have to deal with. Their self-esteem seems highly susceptible to the resistance they receive in response to their commitment to a world that runs on renewable energy.

Finally, the credit restructuring officer, although she carries out a morally complex task, is led in her practical reasoning by many rules and regulations developed inside and outside her organisation. Her job may or may not enable the agent to pursue her ideal of a good life. However, in terms of the agent performing her role, I consider her vulnerability to a diminished self-esteem as low.

In this section, I have presupposed that a negative influence from her social environment on the agent's psychological resources of self-respect and self-esteem would lead to her motivation being hampered. I have reviewed the case material to assess whether agents can be seen to be vulnerable to a diminished self-respect and self-esteem as a result of their social interaction with significant others. The tentative insight we have gained is that committing oneself to help with meeting Humanity's Challenge means that one becomes vulnerable to the potential impairment of one's self-respect and self-esteem as a consequence of the conduct of other agents in reaction to one's commitment. Therefore, these recognitionally secured psychological resources that underpin one's commitment and resolve may be depleted more rapidly when taking a sustainable action compared with taking other morally complex actions.

## 5.6 Overview and diagnosis

Let us take stock of what we have found in this chapter about the motivational challenges for sustainable action in the commitment phase. Commitment has been characterised as a resolution, resulting in felt determination to achieve an end-state that the agent feels determined to achieve. Commitment constitutes a psychological resource for meeting the anticipated challenges of the executive phase. Making a commitment means constructing this psychological resource, which enables the agent to overcome motivational roadblocks that could prevent her from continuing with the action.

Considering the role that intentions play in forming one's commitment, it turns out that the paradigmatic case studies of sustainable actions are the "power grid rebels" (see §3.4) and the purchase manager in the dairy products company (see §3.5). The examples of these agents highlight most clearly the specific motivational difficulties that agents encounter when they make the commitment to take up Humanity's Challenge: they cannot relax on planning; they have to show means-end coherence all the time; and they are highly vulnerable to the belief that they will not succeed with their efforts (intention-belief incompleteness). As these requirements of her commitment could cause the agent to be tempted not to take the sustainable action at her disposal, this means that we have diagnosed three potential motivational challenges for agents who take up Humanity's Challenge.

As regards the relation between commitment and the recognition of agents, we observe that the same case studies are salient. A principle characteristic of sustainable action seems to be that agents have to cope with an at least temporary lack of recognition by other agents and a corresponding change in moral status. The agents involved in the other morally relevant actions do not seem to be affected by the same phenomenon with the same severity. To set oneself on a course of sustainable action means that one needs the psychological resources to withstand the potential impairment of one's autonomy as a consequence of the conduct of other agents in reaction to one's commitment.

I have described the forming of a commitment as the existential moment of binding one's will to follow practical reason and to start executing the proposed action (see §5.2). As the above diagnosis reveals, with regard to intention for sustainable actions it seems that we can tentatively establish that these agents need to rely strongly on their commitment. It seems that agents taking another sort of moral action will more easily find someone else who approves (or disapproves, as the case may be) of what they are doing. A reason for this may be that the ends these agents are pursuing, although perhaps only equally important, seem to be more urgent in the eyes of the public than the ends of the sustainable actions. We can thus enhance our diagnosis of the commitment phase by noting that agents taking a sustainable action need to rely on a strongly developed competency in binding their will to act according to their plan.

Agents who take sustainable action must somehow compensate for the (potential) depletion of their psychological resources of self-respect and self-esteem with a stronger form of commitment. This latter form of commitment they may derive from the ideal that they are contributing to a sustainable future. There seems to exist a kind of commitment that does not particularly arise from our social relations with other agents, but that we obtain by freely and voluntarily deliberating, committing, and doing in the light of an ideal such as meeting Humanity's Challenge. Endeavouring to achieve such an ideal provides the agent with a psychological resource, enabling her to cope with a temporary degrading of her moral status by the social network in which she is acting. In Chapter 7, I will return to this stronger form of commitment. On the basis of the descriptive diagnosis of the Rubicon action phases, I will classify the above-mentioned stronger form of commitment as a second-order commitment (see §7.4). In Chapter 8, I will clarify the second-order commitment with another elaborate case study.

# 6 Motivational challenges in the executive phase

## 6.1 Introduction

In the executive phase we consider an agent who has selected a course of action in the deliberative phase and resolved, in the commitment phase, to carry it out. Despite her all-things-considered judgement in favour of the action and the binding of her will during the existential moment of establishing her commitment, the agent's motivation could still be challenged during the executive phase. Thus, our main questions in this chapter are: Which unanticipated factors might influence the agent's motivation such that she would be hampered in effectively executing her action? And, which differences, if any, can we observe in the case studies between sustainable actions and other actions?

As we are considering complex action situations that extend over time, we may assume that the agent has done some planning towards the end of the commitment phase in order to anticipate potential problems during the executive phase. However, to what extent can we expect the agent to anticipate all possible hurdles? As in the commitment phase, this may depend on intrasubjective factors, as well as on factors that arise from the social interaction with other agents, the so-called intersubjective factors. Both lines of theory are prominent in this chapter.

We consider first a phenomenon that influences the agent in the executive phase – specifically, reconsidering one's intention – by returning to Michael Bratman's theory of intention and plans. Note that the Rubicon model assumes that the agent will return to the deliberative phase when deliberative reconsideration is involved. This happens, for example, when we become aware of new information that puts the reasons in favour of the action into a different light. New features of the action situation may force us to ask: Am I doing the right thing, or do I need to take a renewed decision? Since the agent then returns to the deliberative phase, we focus here on the risk of reconsidering one's intention *without* deliberation, and the requirement to have a robust plan to prevent such nonreflective reconsideration (§6.2).

We continue by discussing some pragmatic complexities of the executive phase. The agent's situation of being dependent on other agents for her effectiveness gives rise to some specific problems in the social interaction with other agents that may result in a detrimental effect on the agent's motivation. These problems relate in part to a lack of competencies on the agent's behalf, and in part to intersubjective factors. In effect, these factors may lead the agent to drop the action, since she becomes aware that her efforts are not successful, despite her strong commitment. As examples of problems that follow from a failure on the agent's behalf, we consider the phenomena of weakness of will and procrastination. We will inquire whether these phenomena affect the agent in a special way when she is attempting to take sustainable action (§6.3).

Then, from empirical psychology we take it that action execution involves the use of resources like self-efficacy and self-control. To what extent does the agent need to draw on these resources when efficacious agency is required in morally complex situations? As already stated, a strong intention and a robust plan do not guarantee successful agency. Should the agent lack the competency of self-efficacy and lack self-control or willpower, her agency could be endangered. Both self-efficacy and self-control seem to work in specific ways when we consider moral action. We will look into the psychological theories of self-efficacy and self-control in order to identify specificities of the sustainable actions (§6.4).

When we engage in the executive phase, carrying out a morally relevant action that extends over time, we display behaviour that aligns with our intentions. As soon as we start acting, we can no longer hide our intentions from other agents with whom we socially cooperate. This, however, means that in the executive phase agents become vulnerable to social pressures that may have an impact on their identity. As in the case of the competencies that form an agent's autonomy (see §5.5), our practical identity is formed in part through our social interaction with other agents. The reactions to our agency by other agents affect the forming of our practical identity. The question to be answered is whether different effects on an agent's identity can be expected as a result of taking sustainable action compared with taking other actions. Since Axel Honneth has spelled out the case of intersubjective identity formation, I return to his theory of recognition and apply it to the case material (§6.5).

Finally, I give an overview of the findings in this chapter with the aim of formulating a descriptive diagnosis of the executive phase (§6.6).

## 6.2 Reconsidering one's intentions and plans

In examining the deliberative phase, we considered the agent who potentially fails to be instrumentally coherent. That is, she believes, roughly, that something is a means to act upon her motive, but refuses to correspondingly intend, or is hampered in so intending (see §4.2). However, now the agent has passed the commitment phase, she has formed a plan of action, and she is in the mode of execution. As we have seen, intentions tend to be stable; they tend to resist reconsideration (see §5.2). This means that we can assume, in our scenario of complicated action situations that extend over time, that the agent will not reconsider her intention immediately, but that any reconsideration involves at least *some* deliberation. However, since we normally do not deliberate about whether to reconsider our intentions, but just go ahead and reconsider (or not reconsider), an important question is: What can be said about the sustainable actions versus the other actions when it comes to reconsidering one's prior intention?

Reconsideration of intentions comes in three forms. I describe them in the following paragraphs in order to make a considered choice as to which form I will apply to the cases. In Bratman's view, reconsideration of one's intention to φ involves seriously reopening the question whether to φ. When reconsidering whether to φ, the intention loses its status as part of the background assumptions against which one deliberates about further intentions. One puts one's prior intention to one side and thus, at least temporarily, one no longer strictly speaking intends to φ (Bratman 1999, 62).

First, consider *nonreflective* reconsideration. This happens when we put our intention aside as a result of "underlying habits, skills, and dispositions" (Bratman 1999, 60). For example, when doing the weekly shopping, I intend to load the vegetables in my shopping cart first, since I am used to the fact that these are placed near the entrance of the shop. Now, suppose the shopkeeper has temporarily put coffee right at the entrance as well: since coffee is on my buying list, I put my intention aside to first collect the vegetables, and take the coffee without deliberation. That is, as a result of some contingent situation in the world, or a habit or disposition of mine, I simply reconsider my intention to take the vegetables first.

The second variety concerns *deliberative* reconsideration. This happens when agents become aware of new information that gives them reason to doubt whether they should continue with executing the current action. This comes to the fore in the example of my intention to switch to solar energy. Suppose that the supplier has gone bankrupt after I have ordered the installation and the new supplier provides me with an offer to execute the original order, but at a 50 per cent higher price. Suppose that the original order was just within my budget, and that the proposal by the new supplier is not. In that case, I will probably enter into the deliberative phase of the Rubicon model again and start considering anew whether I should have solar panels installed.

A third variety of reconsideration of prior intentions is *policy-based.* In this case, I appeal to a general policy of mine to reconsider my prior intention, and I proceed on the basis of this policy without thinking about whether to reconsider my intention. This happens when I base the reconsideration of my prior intention to install the solar panels on, for example, a general policy that prescribes that whenever the commercial conditions under which I am carrying out a plan change, then this gives me a reason to reconsider my original intention (Bratman 1999, 61; Bratman 1989).

As with reconsideration, *non*reconsideration of one's intentions can also come in these three varieties. Yet Bratman argues that there is an asymmetry between reconsideration and nonreconsideration: nonreflective reconsideration of a prior intention will itself be an intentional action, but the same does not hold for nonreconsideration. "Rather, nonreflective nonreconsideration of a prior intention typically consists merely in the *absence* of reconsideration of that intention, an absence to be explained by appeal to relevant general habits and dispositions" (Bratman 1999, 61, emphasis Bratman).

Note that the asymmetry between reconsideration and nonreconsideration coheres with the feeling of determination associated with intention forming during the commitment phase of the Rubicon model. Psychologically speaking, the desired outcome specified by a wish becomes an end state the agent feels committed to achieve. The agent stops her ongoing inner discussion when the intention to realise the wish has been formed. It seems, then, that by "crossing the Rubicon" the agent enters into nonreflective nonreconsideration as a default state of mind. The "relevant general habits and dispositions" to enter into nonreflective nonreconsideration, referred to by Bratman, seem to be well outlined and empirically underpinned.

If we deliberatively reconsider our intention, then the reconsideration itself is a deliberative intentional action. In the Rubicon model we then return to the deliberative phase, in which we explore the various options for acting. Hence, as the reasons for considering the action are again reviewed, the agent can be considered rational in the case of deliberative reconsideration. Similarly, in the case of a policy-based reconsideration we know the reason why the agent is reconsidering her prior intention, and hence her rationality can be exposed. Her policy for reconsideration can be criticised in a discourse with other agents. Obviously, the reflective reconsideration of an intention does not guarantee the agent's efficaciousness, since going back to the deliberative phase also means that she opens herself up to the motivational failures and temptations relevant to that phase (see Chapter 4). Nonreflective reconsideration of a prior intention, however, seems to be an interesting case for our purposes in this chapter (Bratman 1999, 64).

If we nonreflectively reconsider (or do not reconsider) a prior intention, then Bratman considers this the result of a habit or propensity. What can we say about the habits and dispositions that influence our reconsideration of prior intentions? Are there interesting differences on this score between sustainable actions and otherwise morally relevant actions? Is it reasonable for me to nonreflectively reconsider my intention to use my car less over small distances in order to reduce my environmental footprint, just because I am *used* to taking my car? Is it reasonable for me to nonreflectively reconsider my intention to reduce my water consumption in the light of a very likely global scarcity of fresh water by referring to my *habit* of taking a warm shower of half an hour every day?

Bratman develops the following formulation: "The nonreflective (non) reconsideration of a certain prior intention is rational of A if it is the manifestation of general habits of reconsideration that are reasonable of A to have" (Bratman 1999, 66). When looking at the intentions involved in our morally complicated actions, we need thus to ask from a third-person perspective whether the agent's nonreflective reconsideration is the manifestation of general habits of reconsideration that are reasonable of her to have. We are interested in knowing which habits and propensities that potentially influence my plans to execute a morally complicated action it is reasonable of me to have. For example, if my ability to ride a bicycle is underdeveloped, resulting in a feeling of insecurity when cycling through the city, then it is reasonable of me if my habit of avoiding the bike influences my plan to drive less in my car and use the bike instead.

However, there could be other reasons for reconsidering my intention nonreflectively that do not follow from a habit or propensity of mine. First, the effectiveness of my implementing my plan could be seriously affected by unexpected differences in the way the world happens to be, compared with my expectations when I deliberated about my plan. It is perfectly understandable that I put aside many of my prior intentions without any further deliberation when my house has burned down, or when I am hijacked in an aeroplane. Second, there could be a change in my pattern of desires and values, or other intentions of mine could change, causing me to reconsider nonreflectively my original intention to act. I could run into a serious quarrel with my manager at work, or the teacher of my son in primary school could unexpectedly issue a report about his questionable behaviour. In sum, there are potentially many reasons why it could be reasonable of me to reconsider nonreflectively my intention to undertake some morally relevant action (Bratman 1999, 67).

Leaving aside the agent's habits and propensities to unreflectively reconsider prior intentions, we can regard the stability of the agent's plan as an indication of her ability to cope with unexpected events that follow from the contingency in human life. The ability of the agent to plan her actions, trying to anticipate those events that could pose a problem for the execution of the plan, would typically be called practical intelligence. According to Gilbert Ryle:

> an action exhibits intelligence, if, and only if, the agent is thinking what he is doing while he is doing it, and thinking what he is doing in such a manner that he would not do the action so well if he were not thinking what he is doing.
>
> (2000, 29)

A plan is robust when the agent can cope with both smaller and large deviations from the plan in such a way that the end, in the longer term, can still be achieved. Agents are commonly called experienced when their intentions are supported by robust plans that allow them to manoeuvre through the various challenges that could come in their way. Robust plans, an extensive practical experience in similar situations, and a firm intention to achieve the end she has set for herself are crucially important resources for agents attempting to carry out morally complex actions in dynamic environments, where unexpected events may occur all the time.

Moving on, there are two aspects of reconsideration in the light of which we will examine the case studies. The first is the risk of nonreflective reconsideration. As long as the reconsideration is deliberative or policy-based, the agent returns to the deliberative phase, and she changes the plan or abandons it in the light of new reasons that can be defended in a discourse with other rational agents. However, the case we need to consider here – in our discussion of motivational failure in the executive phase – is the *non*reflective reconsideration that potentially influences an agent's course of action, causing her to deviate from her plan without a further thought. The question therefore is: How can we assess the risk that an agent will change her course of action as a result of nonreflective reconsideration? As explained in §5.2, this can be done by estimating the underlying factors that determine the risk: the impact nonreflective reconsideration may have on the agent's effectiveness (severity), and the likelihood that nonreflective reconsideration will occur (probability). The risk that the agent may not follow through with her plan as a consequence of nonreflectively reconsidering her intention will be scored high, medium, or low.

Second, depending on the features of the action and the end to be achieved, there is a requirement for *stability* of the agent's plan, or, in other words, firmness of intention. As we have seen above, depending on the complexity of the action to be carried out and the dynamics of the situation the agent happens to be in, a robust plan expressing the firmness of the agent's intention is required. As before, the important resource of a robust plan and a firm intention can be classified in terms of a 'must', 'should', or 'could have' (see §5.2).

So, let us start with assessing our case studies for the risk of nonreflective reconsideration. As an agent in an institutional role, the credit restructuring officer (see §3.6) will rarely unreflectively reconsider her intention. Since the agent is embedded in a hierarchical bureaucratic organisation, we can assume that her employer approves of her intention to act. Therefore, either the client or another proximate stakeholder will prevent the agent from failing to deliver should she reconsider without a further thought. A potential unreflective reconsideration of her intention will not stay unreflective for long, as it will be brought up as soon as other agents become aware that the agent is deviating from her plan. Hence, although the severity may be medium, the probability is low, resulting in a low risk of unreflective reconsideration.

As we saw with the risk of intention-belief incompleteness, sustainable actions by agents in institutional roles show a different pattern. The purchase manager (see §3.5) seems in an isolated position relative to her immediate environment. The probability that she will unreflectively reconsider her plan therefore is high: there seem to be only negative incentives for her to continue with what she intends. That is, because of her social isolation, her unreflective reconsideration will stay unnoticed by others. And since the agent's efficaciousness entirely depends on her own activity, the severity is also high, leading to a high risk of unreflective reconsideration.

The situation regarding unreflective reconsideration for agents in individual roles could be described as follows. Although the semi-vegetarian (see §3.2) can organise support and assistance for her plan, we should realise that in high-income OECD countries vegetarians typically make up less than 10 per cent of the population.[1] This means that the probability is high that semi-vegetarians, for reasons of social convention, cannot stick to their intentions. The agent faces numerous situations in which she cannot live up to her intention to be a semi-vegetarian, which may tempt her to put her intention aside. While the severity of deviating is at least medium, the risk is high that the agent will change her intention unreflectively.

For the doctor without borders (see §3.3) the risk of unreflective reconsideration is probably low. As noted earlier in relation to means-end coherence of the agent's plans (see §5.3), his alternatives for action are mainly under the control of significant other agents. As soon as he has decided to join them and has passed the application procedure, the MSF organisation will organise his mission to an area in need of emergency aid. Once he is on mission, unreflective reconsideration is unlikely to happen. Hence, although severity may be medium, probability is low, and the risk of unreflective reconsideration is therefore low.

Lastly, the "power grid rebels" (see §3.4) face a situation that is similar to that of the purchase manager (see above). At the start of their project, negative incentives for pursuing their plan seem to dominate. From the start, these agents organise their project in their private time, on top off their other engagements (family life, jobs, hobbies). Therefore, the probability that the turmoil of their daily routine will cause them unreflectively to put their project on hold for some time seems high. As discussed in relation to intention-belief incompleteness, severity is also high, since their project entirely depends on them; hence unreflectively slowing down the project will almost immediately lead to inefficaciousness. I thus consider the risk of unreflective reconsideration to be high.

When considering the requirement for firmness of intention, the cases can be split into three sustainable action situations and two other morally complicated situations. While executing their decision, the credit restructuring officer and the doctor without borders are strongly interacting with their social environment, in particular with those people who will be affected. This means that the firmness of their intention is mainly a function of the support they obtain for their decision. The robustness of their plan does not necessarily have to be established by these agents *themselves*, as they can rely on thorough support in this regard. Therefore, I determine the requirement for firmness of intention in each of these cases as a 'could have', meaning that firmness of intention is desirable but not absolutely necessary for the agent to be efficacious. In contrast, as our considerations so far have shown, for the agents of the sustainable actions, firm intentions and stable plans to reach the ends they have decided upon are a 'must have'.

In this section, we have considered the agent and the risk of unreflectively changing one's intentions. I have diagnosed that this risk is high for the paradigmatic agents compared with the other agents (i.e. the "power grid rebels" and the purchase manager vs the other agents). The former could form intentions and plans, and then refrain from executing, without anyone noticing. In order to withstand the resistance they encounter from other agents, they need to stand firmly for their intention by having a robust plan.

## 6.3 Social cooperation and philosophy of action

Up to this point in the phases of the Rubicon model, the agent has succeeded in choosing a proposal for action in the deliberative phase, and she has decided during the commitment phase that she intends to execute this action. However, even if the agent has deliberated thoroughly, her commitment is strong, and her plan robust, her acting towards the goal may fail as a result of her motivation being increasingly hampered. That is, her motivation starts to fade away, not because she has gone into nonreflective reconsideration of her intention (see §6.2), but because her perception is that she does not seem to be coming closer to the end she has set. In one way or another, the agent becomes aware that her intention does not seem strong enough to meet the unanticipated challenges she is confronted with in the executive phase. Consequently, she is tempted to give up, not because she believes that the goal of her action is no longer worth striving for, but because the agent believes, at least temporarily, that she will not succeed despite her effort. The agent gives up on acting towards the goal she has set, because she simply no longer sees the possibility of reaching the goal of her action. In short, the agent considers pursuing the action to be useless.

Note that this is not a form of the agent believing that she is not capable of following through (i.e. intention-belief incompleteness, see §5.4). That phenomenon takes place when the agent is agnostic about whether she will succeed when she starts to put her plan into action. For example, this happens in situations where the agent is aware that she may be ill-equipped to handle the difficulties that come with executing her action. In the situation under consideration here, however, the agent feels capable of executing. She may, however, be prevented from progressing because of certain intrapersonal failures on her part, of which she may not be aware. Or, she may lack certain psychological and physical resources that she needs specifically in the executive phase. Or, she may become frustrated following recognitional problems that come about in the social interaction with other agents (see §5.5).

Now, by elaborating the situation of the agent as I have done, the immediate objection could be made that I am opening a new debate about the question of organising one's actions so as to achieve one's ends in the most efficient and effective way. This could bring us into the territory of other academic disciplines, such as business administration, economics, sociology, and political sciences, where I do not wish us to go. Instead, I will focus on discussing the above-mentioned three kinds of causes of motivational failure. First, later in this section, I will discuss phenomena from the philosophy of action, such as weakness of will and procrastination, which have an indirect influence on the agent's motivation. The detrimental effects on the agent's motivation result from the reaction of other agents to her intrapersonal failures. Second, I will discuss diminished motivation that is due to a lack of self-efficacy or self-control. These are well-known resources, mainly studied in psychology, without which agency becomes extremely difficult (§6.4). Third, as in Chapter 5, I will discuss the motivational difficulties that arise when agents must act in an environment in which there are other agents who do not recognise them, for example as a professional colleague. I am interested in the detrimental effects on an agent's motivation of these kinds of flaws in the social cooperation between agents (§6.5).

However, in order to prepare for the further conceptual-phenomenological analysis in this chapter, I will start with some thoughts about what it means to execute an action in an environment where social cooperation with other agents is necessary if the agent is to be in any way effective. These reflections on action execution are important for understanding why it is even necessary, in the following sections, that motivational problems be considered. Thereafter, I will continue with discussing weakness of will and procrastination. These phenomena are intrapersonal, and the agent may be aware (or not) that her ineffectiveness is caused by her own deficiencies. I will argue that agents in performing a sustainable action face more severe problems with weakness of will and procrastination than do the other agents.

When agents intend to execute their action in the morally complicated situations we consider in this book, they need to make an utterance, typically involving language, to set things in motion. These agents must mobilise other people, so as to organise and encourage them to act in a concerted way, in order to bring about a particular objective. As I have said above, I will not discuss ways of organising one's environment in the most effective and efficient way, nor will I discuss the vast terrain of competencies that people may need to be an effective agent. Instead, I will focus on the element of involving other people by means of sharing one's plans with them and being in some way dependent on them for achieving the end of one's action. This means that I want to investigate some specific problems arising in the social interaction with other agents in the executive phase that may have a detrimental effect on the agent's motivation.

An interesting new feature of this phase, at least for the cases of moral action under consideration here, is that the agent confronts her social environment with her commitment to act, followed by her behaviour aimed at realising the intended goal. Therefore, the agent's efficaciousness in this phase depends to a considerable extent on her capacity to mobilise her social environment to support her in acting towards the goal she wants to achieve. This means that executing the action efficaciously depends to a high degree on the pragmatic skills of the agent. These skills are related to communicating with other agents, such that the effect of the communication on other agents is that they start contributing towards achieving the intended goal. The pragmatics of action execution thus relies on the agent's communicative skills in terms of creating support for her plans among other agents.[2]

In expressing her commitment to the action, the agent potentially causes a difference between herself and her immediate social environment. In the morally complicated action situations under consideration here, we assume that other agents do not necessarily share the same commitment as the initiating agent. At the same time, the initiating agent depends for her execution on these other agents, in the sense that their actions are necessary for her to succeed. Since the initiating agent in her execution depends on other agents' execution, a necessary condition is that the other agents find a motivating reason to help the initiating agent with her action. For the other agents to start contributing to the execution, there must be a reason that explains their behaviour, which is a motivating reason.

When the other agents begin their execution, there is at least one assumption they will make. They may have a variety of strategic reasons of their own that explain their behaviour; however, they assume that their action is instrumentally coherent with the end the initiating agent has set (see §4.2). The interaction between the initiating agent and her social environment, therefore, shows the mutual interdependence between the following phenomena: the expression of the initiating agent's commitment, the reaction to that expression by other agents, the forming of motivating reasons in the other agents, and the expectation of the other agents about the contribution of their action to the initiating agent's overall plan.

In particular, the other agents expect the initiating agent to follow through on her commitment, unless they have the same understanding of the reasons why the initiating agent has stopped. Since these are moral actions, for which the initiating agent can be held accountable, she cannot delegate fully to other agents. The results of the execution by the other agents will need to be integrated within the overall plan so as to realise the end of the initiating agent's action. This means that other agents expect the initiating agent to show behaviour during the executive phase that expresses her continued commitment, or, in other words, that she is still motivated. When the initiating agent shows signs of a lack of commitment, the other agents may lose their motivation to contribute, which in turn may render the initiating agent ineffective.

However, the initiating agent's motivation may fade away due to a lack of support from the other agents, which in turn is caused by their reaction to the initiating agent's executive failures, of which she may not be aware. This could happen when the initiating agent suffers from typical executive failures, such as weakness of will or procrastination. It thus makes sense to consider the question to what extent these phenomena from the philosophy of action matter in the case studies, since the reaction of the other agents to the initiating agent being exposed to these phenomena could be that they withdraw their support, which in turn may demotivate the agent. Thus, the occurrence of weakness of will or procrastination in the initiating agent could lead to a reactive attitude in other agents that makes them end their support, which in turn leads to a diminished motivation in the initiating agent. I will now discuss these phenomena from the philosophy of action.

The proposal for action that came out of the deliberative phase has survived the commitment phase; that is, the agent now acts in conformity with her intention to realise the desired end. Importantly, this means that the action is not a mere happening, is not coerced, and is intentional, as opposed to mistake, accident, or inadvertence (Campbell 2000, 107). As explained above, by expressing her commitment, the agent changes the social interaction with other agents, and at the same time she depends on these other agents for success in executing her action. If the agent, then, in some way fails to live up to her own commitment, she will be vulnerable to criticism from other agents, who expect her to follow through. As a result, the other agents may withdraw the support on which the initiating agent depends. A failure to deliver on her commitment happens if the agent displays weakness of will or procrastination. I will discuss these concepts in order to find out the role they play in our descriptive diagnosis of the executive phase.[3]

In order to understand *weakness of will*, let us consider first *akrasia*, standardly understood as the "(free, intentional) action contrary to one's better judgment. … Rather than the two – action and judgment – being in concert, there is a dissonance or lack of correspondence between the two that marks off the action as akratic" (Stroud 2010, 55). Sarah Stroud has recently introduced a distinction between akrasia and weakness of will by stating that the latter "is a species of irresoluteness, or failure to follow through on your intentions. It is a failure to do what you have decided you will do – a failure to stick to your plans" (Stroud 2010, 60).[4] Following the theory of intention, there is no reconsideration of intention taking place here (see §6.2). The agent still intends to act, but simply does not follow through. As I have argued, the reaction of other agents to a weak-willed initiating agent could be that they end their support, which in turn could have a detrimental effect on the agent's motivation. Hence, for our investigation an important question is whether we can make relevant distinctions between the case studies in terms of this phenomenon.

Note that the case material in Chapter 3 consists of examples of moral action. These are actions that can be criticised by those who are affected, due to the moral implications the action has for them. What is the likely effect that an agent's failure to stick to her plans will have in such an action situation? In the first instance there seems to be a difference between the case of the credit restructuring officer (see §3.6) and the other cases. For the former agent there is little possibility not to follow up on her plans, since those who are affected monitor her actions closely. Should she not follow up on her plans, this will be noticed quickly, and the agent will be reminded of her inaction. As the beneficiaries of the actions by the other case study agents are future generations or people who are far away, and as Scheffler's phenomenology of agency explains that "near effects of our actions have primacy over remote effects", it seems that the agent can be weak-willed almost indefinitely. This is what can be noted about the *start* of the action execution – that is, the action initiation.

Once execution has *begun*, however, it appears that the doctor without borders (see §3.3) differs from the other (sustainable) cases. The doctor will be criticised immediately in the event that he does not deliver the aid his patients expect from him. Once he is on mission, there is hardly any need to reaffirm his intention, since the local situation has become part of his moral horizon. Quite literally, he stands in front of the beneficiaries of his action. In other words, the doctor's action situation makes reference to affected people who have an interest in the agent following through, and who rely on him in this respect. However, for the agents taking a sustainable action, especially the "power grid rebels" (see §3.4) and the purchase manager (see §3.5), it seems that they have to reaffirm their intention repeatedly. These agents operate in social environments that are hostile to their plans. Weakness of will as the agents display it will then even be welcomed, as it is in the interests of the resisting agents that the action be delayed or cancelled.

According to a similar formulation by Holton and McIntyre, "weakness of will is unreasonable revision of a contrary inclination defeating intention (a resolution) in response to pressure of those very inclinations" (Holton 2009, 78). In my view, the same argument as for Stroud's definition applies. The credit restructuring officer (see §3.6) will hardly be in the position to revise a resolution, for the reason that her behaviour is under scrutiny by those affected. The doctor without borders (see §3.3), as long as he is not on mission, can unreasonably revise his resolution for as long as he wants. However, once he is on mission, this form of weakness of will is also ruled out. Among the agents taking sustainable actions, there are many possibilities for each agent to unreasonably revise her resolution in response to inclinations that she anticipated and expressly formed her intention in order to defeat (Stroud 2010, 62). Thus, displaying weakness of will is likely to occur when agents act sustainably, which may result in other agents withdrawing their support. This in turn may lead to the agent's motivation fading away. I conclude, therefore, that weakness of will seems to help us in getting a clearer picture of the motivational difficulties of taking sustainable action.

Finally in this section, I analyse what *procrastination* does to agents taking sustainable actions compared with other actions. From the outset it is obvious that as agents, due to our physical limitations and constraints on possibilities of action within our environment, we cannot do everything we intend at once (Stroud 2010, 53). We need to plan our actions in time, and, hence, we require a definition of procrastination that is more specific than saying merely that agents put off some things as a result of ordering their activities. Procrastination should be understood not just as putting off, but also as a failure to do something (Stroud 2010, 59).

It seems that Joel Anderson's definition of procrastination as "culpably unwarranted delay" captures this failure adequately (Anderson 2016, 47). As with weakness of will, delaying a task can happen in all cases we consider here, with the same consequences: because it may result in other people running into problems, the agent will be criticised if she delays. For the affected persons, it does not matter whether weakness of will or procrastination of the agent causes the delay; they blame her anyway. However, if the agent can convince the affected people that they will in fact be better off if she delays the task, then she is strategically delaying, not procrastinating (Anderson 2016, 48). Hence, we assume that the delay is not strategic, and that it is problematic in the eyes of those who are affected.

In order to count as procrastination, the agent *herself* must have a sense that delaying the task is unwarranted (Anderson 2016, 49). This means that we are considering agents who sense that their delaying is unwarranted and expect criticism for this delay by the affected people. Just as with weakness of will, before going on mission the doctor without borders can procrastinate without being criticised. However, as argued above, as soon as he is on mission, there will be hardly any room for procrastination. Because in sustainable action situations agents act in the interest of future generations, who cannot express their criticism yet, the risk of procrastination in these cases seems higher than in cases where current people have issues with the delay. When executing their action, agents acting sustainably are thus more vulnerable to culpably unwarranted delay than the other agents. Future generations are remote and do not enter the agent's moral horizon unless she exercises strength of will.[5]

The above considerations make clear that weakness of will and procrastination can result in reactions by other agents on whom the initiating agent depends that in turn may lead to diminished motivation in the initiating agent. These phenomena thus constitute motivational challenges for agents taking sustainable action. Agents taking sustainable action seem more likely to display weakness of will and procrastination than the typical altruistic case of the doctor without borders and the case of the credit restructuring officer, who faces moral dilemmas all the time.

## 6.4 Self-efficacy and self-control

I have argued that other agents expect the initiating agent to show behaviour during the executive phase that expresses her commitment, or, in other words, that she is motivated. In other words, when the initiating agent shows signs of a lack of commitment, the other agents may lose their motivation to contribute, which in turn may render the initiating agent ineffective. As a consequence, the initiating agent's motivation diminishes due to a lack of support from the other agents (see §6.3).

However, in order to stay committed and deal with the unanticipated challenges of the executive phase, the agent requires psychological and physical resources. In this section, I discuss two potential obstacles, investigated in empirical psychology, which could lead to a lack of resources that are required for committed agency: problems with self-efficacy and self-control. Under self-efficacy one could subsume those agential competencies that enhance the agent's control over events that affect her life. I will consider the theory of Albert Bandura that details the resources of self-efficacy that agents need to draw on. The second potential obstacle relates to self-control resources or willpower. From daily experience we know that this resource is limited. Therefore, given the obstacles to be dealt with, from the outset it seems that a conscious allocation of self-control is necessary for efficacious agency. I will consider the theory of willpower as developed by Roy Baumeister and others.

Recall that the Rubicon model sets out four conditions that have to be fulfilled for a goal intention to lead to the initiation of appropriate actions (see §2.5): first, the volitional strength of the goal intention has to be large enough as compared with that of other competing goal intentions; second, the situation has to be favourable in that the particular goal intention can be readily initiated as compared with initiating competing goal intentions; third, the situation has to be favourable in that the particular goal intention can be readily initiated as compared with relevant future opportunities one hopes to encounter; and, fourth, the goal intention should have an additional advantage over competing goal intentions, assuming that the opportunity specified by the initiation intention is present. It follows from our discussion of the strength of an agent's intentions (see §6.2) that satisfaction of the first of these conditions we can inherit from the commitment phase. The other conditions will have to be met when the agent proceeds through the executive phase.

Meeting these conditions means that a necessary competency for agents to succeed with their action is that they are able to exercise control over events that affect their lives. By exerting influence in spheres over which they can command some control, agents are better able to realise desired futures and to prevent undesired ones (Bandura 1995, 1). What makes the execution of an action more or less difficult is the degree to which the agent can be causally effective in realising the desired result. We will now look into some features of an agent's competency of self-efficacy that are relevant for our consideration of the case studies and then proceed with some remarks about the sources of an agent's self-efficacy beliefs as investigated in psychology.

One can look at an agent's self-efficacy from two equally important angles: on the one hand, we can consider the agent's competencies irrespective of characteristics of the action situation; on the other hand, we can look at features of the agent's environment, whilst disregarding her competencies for successful agency. If we take the latter angle, we are interested in the structure of the action situation and whether this gives guidance in identifying the specific difficulties of sustainable action. Focusing on the agent's environment, our goal is then to distinguish between actions that can be initiated and executed entirely by the agent herself, and actions where the agent can only exercise a marginal influence on whether she succeeds. In cases of the latter sort, the agent's self-efficacy thus depends largely on the environment in which she operates, and to a much lesser extent on the agent's competencies for executing the action.

Looking at self-efficacy from the former angle, we aim to study the agent's competencies for organising the different elements of a complex activity so as to enable them to work together effectively. Independent from the features of the action situation, the competency of self-efficacy, understood as coordination of other people or institutions required to perform the action, can vary considerably. The competency to coordinate could also be described as the competency to motivate other agents to engage in what the agent intends to achieve. Being a competent coordinator, or leader, of the actions by other agents means that one has acquired the capabilities needed to get things done under various, and often difficult, circumstances. Without entering into a discussion about theories of leadership, one can say that competent leaders have certain capabilities related to rhetoric, persuasiveness, and pragmatics.

Thus, recognising the conditions that have to be fulfilled for a goal intention to lead to the initiation of appropriate actions depends on features of the action situation and on the agent's competencies of coordination and leadership. Depending on their experience, agents may be more or less skilled in correctly determining whether a given situation is favourable for implementing their intention or not.

The psychological theory of self-efficacy tells us that an agent's level of motivation, as it is related to her self-efficacy, is "based more on what they believe than on what is objectively the case" (Bandura 1995, 2). Hence, this theory focuses on how an agent's belief in her causal capabilities – that is, her self-efficacy belief – can be explained. Such perceived self-efficacy refers to beliefs in one's capabilities to organise and execute the courses of action required to achieve the desired goal. Bandura outlines four sources of an agent's efficacy beliefs, which I will summarise here in order to apply them to the case material later (Bandura 1995, 3–5).

First, and most influential, are *mastery experiences*. These are successes in similar action situations encountered by the agent in the past. Mastery experiences provide the most authentic evidence of whether one is capable of efficacious agency in a new action situation. Successes build a robust belief in one's personal efficacy; failures undermine it. Developing the appropriate mastery experiences involves acquiring the cognitive, behavioural, and self-regulatory tools for establishing and executing courses of action under changing circumstances in one's life. If people experience only easy successes, they come to expect quick results and are easily discouraged by failure. A strong sense of self-efficacy requires experience in overcoming obstacles through one's persistence.

Second are *vicarious experiences*. These are similar actions successfully executed by social role models. Seeing people similar to oneself succeed by persistent effort raises the agent's belief that she, too, possesses the capabilities to carry out a similar action. By the same token, observing others fail despite their efforts lowers the agent's expectation about her own efficacy. Social role models do more than simply provide a standard against which to judge one's own capabilities. Agents seek competent role models that possess the competencies to which they aspire. Acquiring such improved competencies raises the perceived self-efficacy of the agent.

Third is *social persuasion* as a way of strengthening people's beliefs that they have what it takes to succeed. Agents who are persuaded verbally that they possess the capabilities to carry out specific actions are likely to mobilise and sustain greater effort than those operating under the influence of self-doubts, who focus on personal deficiencies when problems arise. Self-efficacy is also easily undermined by social persuasion, as unrealistic expectations are quickly disconfirmed by the disappointing results of one's efforts. People who have been persuaded that they lack capabilities tend to avoid challenging activities, and they give up quickly in the face of difficulties.

Fourth, agents also rely partly on their *physiological and emotional states* in judging their capabilities. They interpret their stress reactions and tension as signs of vulnerability to poor performance. In activities involving strength and stamina, people judge their fatigue, aches, and pains as signs of physical weakness. Mood also affects people's judgements of their personal efficacy. Thus, agents can alter their efficacy beliefs by enhancing their physical status, and by reducing stress and negative emotions.

I will now review the cases from Chapter 3 with regard to the four sources of self-efficacy beliefs listed above. Again, we are looking for differentiating characteristics of sustainable action, as compared with the other cases.

The (semi-) vegetarian (see §3.2) finds herself in a situation where she can decide what to eat and not to eat. Although her intention needs to be firm, and although she may reconsider unreflectively many times (see §6.2), she can unilaterally implement her decision. However, assuming that this is the first time that the agent has decided on a low-meat diet, she will not have any experience that supports her belief in self-efficacy. She has no mastery experiences. On the other hand, as there are quite a few social role models following a vegetarian diet, the agent can rely on vicarious experiences. As to social persuasion, given that vegetarians are still a minority of less than 10 per cent in high-income OECD countries (see §6.2), there will perhaps be support from her family, but outside this small circle hardly any social support can be expected. Finally, and quite literally, I assume that the agent's physical and emotional states inform her about her self-efficacy.

According to the case description, a doctor without borders (see §3.3) works under extreme conditions that require different competencies than needed for working in his affluent home country. The surgeon is experienced under the conditions in his home country; he has no experience in war zones in developing countries. Thus, his mastery experiences are mixed: in order for the US and not present for Nigeria. Just as with the vegetarian, the doctor is not the first one to take this decision. There will be social role models that support his choice (i.e. vicarious experiences). Also, social persuasion will be strong, since his friends, his family, and the general public will admire him for pursuing this change. However, as doctors for MSF work under extreme conditions, the surgeon may interpret his stress reactions and tension as signs of vulnerability to poor performance. That is, his physical and emotional states may not support him.

The "power grid rebels" (see §3.4) are in a somewhat different situation. Their self-efficacy is complex and relies on many factors that are outside their control. Although they have certain means at their disposal (e.g. the right to organise a referendum), these are very dependent on the voting behaviour of the local citizens. The complexity of the coordination in this case is high, as the agents have to overcome various serious obstacles that could put their initiative in jeopardy. Their project is unique in the sense that there are no mastery experiences available. Although there may be a few relevant sustainable action role models, vicarious experiences will hardly be available. As we have seen, there is hardly any external pressure on these agents. Social persuasion from outside their small group is nil at the start of their project, only increasing later on when fundraising turns out to be successful. Finally, the "power grid rebels" will likely suffer from stamina problems, strains, and other discomforts, due to situations in their unique project.

When we move from the private to the professional sphere, the level of an agent's self-efficacy changes. In the example of the purchase manager (see §3.5), we observe that the network of hierarchical relationships in the company influences her effectiveness significantly. Basically, her low level of self-efficacy prevents her from making progress with her action, until the company's CEO supports her initiative. The level of complexity of her project, at the same time, is high. As with the "power grid rebels", there are no previous experiences on which the purchase manager can rely to support her self-efficacy belief. Also, there are no vicarious experiences available. Social persuasion is not simply absent but even negative during the first phase of her project. Only when the CEO engages with the action does this become a source of efficacy belief. Again, just as with the "power grid rebels", the purchase manager needs good mental and physical health to withstand the tensions and pains she faces.

The case of a credit restructuring officer in a bank (see §3.6) looks different. Although agents in these roles typically have to submit their proposal for a particular client to managers higher up in the organisation, they have ample opportunities to influence the decision by selecting and judging information. They have reached their level of seniority by showing their competencies in similar situations. Such agents can therefore rely on mastery experiences. Likewise, there will be senior (former) colleagues who act as social role models, providing vicarious experiences. And, since the officer has been hired to perform this role, there is more social persuasion than in the other cases. The job, though, is very demanding, and the agent may suffer from stamina problems and headaches. All in all, this agent needs good physical and mental health, and she can rely on the other sources for her self-efficacy belief.

Overall, I observe that the "power grid rebels" and the purchase manager are again paradigmatic for taking sustainable action in that these agents can only rely on good physical and mental health for support of their self-efficacy belief. Since both agents were able to manage their tasks until completion, it seems an interesting question which source of motivation was available to them that made them capable of overcoming the motivational obstacles we have investigated so far (and there are more to come, see §6.5). However, for some further insights from psychology, I will continue with looking into self-control as a source of motivation.

So far, I have diagnosed that agents taking sustainable action seem to face particular difficulties with continuing to be motivated throughout the executive phase. We learn from the commitment phase that they must have a robust plan; they have to show means-end coherence all the time; and they are highly vulnerable to the belief that they will not succeed with their efforts (intention-belief incompleteness). Concerning their commitment and the way they are recognised by other agents, we observe that these agents have to cope with an at least temporary lack of recognition and corresponding change in moral status. To set oneself on a course of sustainable action means that one needs the inner resources to withstand the potential impairment of one's autonomy as a consequence of the conduct of other agents. We have seen that in the executive phase these agents seem more likely to display weakness of will and procrastination. In comparison to other action situations, they are more prone to unreflectively reconsidering their intentions, and they have a strong need for a robust plan in order to succeed. For these agents to be able to overcome these obstacles, we can say that their commitment must be strong indeed.

At the same time, we tentatively observe that the paradigmatic agents taking sustainable action cannot rely on sources on which agents taking other moral actions can draw for their self-efficacy belief. They cannot draw on mastery experiences, nor on vicarious experiences or social pressure. A question, then, is: How can the paradigmatic agents overcome these significant motivational hurdles? This question is important, as answering it may help us in finding ways to facilitate agents in taking sustainable action (e.g. by means of an institutional policy).

An obvious candidate for an answer could be that the agents who are successful in staying on course in executing their intended sustainable action possess more self-control (known colloquially as willpower) than other agents. We will thus look into the psychological theory of self-control to see whether this theory may help us with this question. Michael Inzlicht and Brandon J. Schmeichel propose to understand self-control in terms of the mental capacity of individuals to override or alter their own thoughts, emotions, and behaviours. The theory of self-control tells us that this mental capacity can be depleted, and the task is to explain the causes thereof. What causes acts of self-control at $t_1$ to reduce performance on subsequent, seemingly unrelated self-control tasks at $t_2$ (Inzlicht and Schmeichel 2012, 450)? For example, it turns out that people who have recently exerted self-control respond more aggressively to insults, suggesting a reduced capacity to regulate aggression (Inzlicht and Schmeichel 2012, 452).

Roy Baumeister and others have developed a model of self-control as drawing on a resource that is used when people act. This means that agents, while exerting self-control, use up a certain amount of energy, which results in failing to resist temptations if this source of energy is depleted. If agents try to resist temptation, they may find themselves feeling forbidden desires more strongly, just when their ability to resist them is low. Using up the resource, which Baumeister calls ego depletion, thus creates a double whammy: one's self-control is diminished and one's cravings feel stronger than ever (Baumeister and Tierney 2012, 31). Since its publication, the resource model has been tested many times. Although it has been challenged (see below), the resource model brings Baumeister and John Tierney to conclude, first, that agents have a finite amount of self-control that becomes depleted as they use it; and, second, agents use the same stock of self-control for all sorts of tasks. Thus, the same supply of willpower is used to cope with frustrating traffic, writing an academic paper, dealing with one's demanding manager, and doing the weekly shopping (Baumeister and Tierney 2012, 35–36).

Inzlicht and Schmeichel challenge the resource model. The question they raise is: What is the resource in Baumeister's resource model? Is it merely a metaphor, or should we stick to the literal interpretation, leading to the frequent suggestion that it is glucose in the bloodstream, which is depleted between $t_1$ and $t_2$? First, Inzlicht and Schmeichel propose that engaging in control at $t_1$ may lower the *motivation* to engage in control at $t_2$ and shift motivation away from further restraint towards engaging in something more gratifying. That is, what happens between $t_1$ and $t_2$ "is that people become unmotivated to regulate themselves; it is not that they cannot regulate but that they choose not to regulate" (Inzlicht and Schmeichel 2012, 454). As an explanation for this phenomenon, Inzlicht and Schmeichel suggest that self-control can be construed as a competition between two opposing forces: the force that motivates the expression of an impulse (i.e. "impulse strength") versus the countervailing force that overrides the impulse (i.e. "self-control strength"). As they state:

> In this view, self-control succeeds when the impulse is relatively weak, when control is relatively strong, or some combination of both of these factors. Conversely, failures of self-control may stem from strong impulses, weak control, or a combination of both factors.
>
> (Inzlicht and Schmeichel 2012, 455)

Second, Inzlicht and Schmeichel propose that the shift in people's motivation is accompanied by a parallel shift in *attention* away from cues signalling the need to control and towards cues signalling the possibility of reward. That is, self-control failure at $t_2$ may come about because people fail to notice when control is actually needed.

The relevance of this empirical research is that it may help us make a reasonable conjecture as to why the agents engaged in sustainable action manage to overcome the significant motivational hurdles, while at the same time lacking the sources of their self-efficacy belief. To me, it seems that the paradigmatic agents derive their self-efficacy belief from their commitment to contribute to meeting Humanity's Challenge. As I suggested, these agents must somehow compensate for the (potential) depletion of the psychological resources of self-respect and self-esteem by constructing another kind of resource (see §5.6). In that sense, there seems to exist a kind of commitment to work on meeting Humanity's Challenge that renders these agents self-efficacious. They do well, in that they have found a source of their self-efficacy belief, which is their commitment to endeavouring towards the good, through working on meeting Humanity's Challenge.

This suggestion seems compatible with Baumeister and Tierney, who explain how self-control can be improved. They emphasise the importance of setting a clear goal. Just as I have specified it (see §1.2), this means that the agent performs self-control by guiding her actions towards a specific, intentional, and meaningful purpose (Baumeister and Tierney 2012, 62). By elaborating the example of Henry Morton Stanley, they draw attention to the importance of forming habits that strengthen self-control. In their view, the start of forming a habit is being precommitted, a strategy that Stanley used with great success. Baumeister and Tierney formulate this strategy as follows:

> The essence of this strategy is to lock yourself into a virtuous path. You recognise that you'll face terrible temptations to stray from the path, and that your willpower will weaken. So you make it impossible – or somehow unthinkably disgraceful or sinful – to leave the path. Precommitment is what Odysseus and his men used to get past the deadly songs of the Sirens. He had himself lashed to the mast with orders not to be untied no matter how much he pleaded to be freed to go to the Sirens.
>
> (Baumeister and Tierney 2012, 151)

Richard Holton argues for a similar strategy for resisting temptation. The self-control the agents taking sustainable action perform is that they resist reconsidering their intention. Their precommitment has become a habit that enables them to resist temptations to which they are prone. I understand precommitment in the sense Holton understands a resolution: "intentions formed by the agent with the very role of defeating any contrary inclinations that might emerge" (Holton 2009, 119).

In terms of the theory of Inzlicht and Schmeichel: being committed to work on Humanity's Challenge becomes an impulse for the agent to act upon when her self-control is depleted as a result of acting on other tasks. That is, the "power grid rebels" and the purchase manager, after having completed different tasks on which they depleted their self-control, obtain a self-control boost by turning to work on their goals of a sustainable future. Working towards their sustainable goal does not seem to use up self-control but seems gratifying and rewarding to these agents. The shifts in motivation and attention that occur when they deplete self-control on other tasks seem to take place towards the goals that give these agents a meaning in life. They do not seem to obtain a reward from the impulse to eat chocolate bars (say), but from continuing with their acting towards the goal of meeting Humanity's Challenge. Their self-control is enhanced instead of depleted by taking sustainable action. Their commitment to meet Humanity's Challenge keeps them going.

In this section, I have considered the phenomena of self-efficacy and self-control as psychological and physical resources on which the agent must rely so as to be effective, especially when responding to unanticipated challenges. Other agents, on whom the initiating agent depends for executing her action, presuppose that she has these resources at her disposal. As the analysis of the cases shows, however, this does not seem self-evident. The paradigmatic agents taking sustainable action can rely on only one of the sources of self-efficacy belief: good physical and mental health. By looking at theories of self-control, I conjecture that their commitment to work on Humanity's Challenge forms a motivational resource the paradigmatic agents can rely on, which could explain their efficaciousness despite how little support they receive from other agents.

In Chapter 7, we will further elaborate the kind of commitment that is at stake here, functioning as it seems to do as a source of the agents' self-control and their self-efficacy belief. I will propose a so-called second-order commitment as a kind of commitment from which agents receive a boost in self-control and a strengthening of their self-efficacy belief (§7.4). In Chapter 8, we will discuss in detail the case of an agent who is committed to implementing a sustainable institutional policy. This agent faces motivational difficulties in the form of severe inaction or stalemate, which she can only prevent through changing the relationships with the parties involved in the action situation. We need, however, to consider one more motivational challenge in the executive phase, to which we proceed in the next section.

## 6.5 Sustainable action and practical identity

The social interaction with other agents belongs to pragmatic complexities of the executive phase. This phase can be characterised by social interactions between the agent and other agents on which she depends, which become hard to unravel when they start affecting the agent's motivation (see §6.3). As in the case of the competencies that form an agent's autonomy (see §5.5), our practical identity is formed in part through our social interaction with other agents. In this section, we consider the constitution of the agent's practical identity through social cooperation. The question to be answered is whether different effects on an agent's identity can be expected as a result of taking sustainable action, when compared with other actions. The agent's commitment, and therefore her motivation, may be affected by recognitional patterns in her interaction with other agents. We will look first into the theory of intersubjective recognition by Axel Honneth, and then we will apply some of the concepts to the case material.

Honneth's recognitional account of identity aims to clarify how an agent's identity emerges while she is in social interaction with other agents, and also as a result of such interaction. Honneth argues that the development of a subject's personal identity presupposes, in principle, certain types of recognition from other subjects, and that the subjects in such recognitional relationships have the obligation of reciprocity to each other (Honneth 1995, 37). However, before a process of intersubjective identity formation can be described, the realm in the human mind where our identity resides should be indicated. This realm is named the psychical.[6]

While we are taking actions in social cooperation with other agents, we become aware of the psychical. This happens when we have the experience of ourselves, for example when we are prevented from carrying out an action as intended. By acting and by experiencing the necessity to adjust one's interpretation of a practical situation, as a result of one having to accommodate different outcomes than expected, one becomes aware of one's subjectivity, and thereby the realm of the psychical is opened up (Honneth 1995, 72).

Drawing on the social psychology of George Herbert Mead, Honneth distinguishes between the 'me' and the 'I' within what Mead calls the "psychical realm". The 'me' is to be understood as the reflection of the other's image of me, as far as I can reproduce in my mind the other's reaction to my action. The 'I', on the other hand, represents the source from which all of my current actions spring in a creative response to others with whom I interact (Honneth 1995, 74). A person's objective 'me' arises in the communication with others following two equally important processes: on the one hand human self-consciousness is formed by calling up in oneself the way in which one's action affects others; and on the other hand the perception of one's action by others is a prerequisite for the development of one's self-consciousness. On Mead's account, the 'I' seems to exist by implication, because of its continuous dialogue with the 'me'. However, the 'I' remains behind the scenes, and, according to Mead, can never exist as an object in consciousness. The 'me', on the other hand, is formed by the cognitive predictions that others have of my behaviour, and, more importantly, also by their normative expectations of me (Honneth 1995, 76).

According to Honneth and Mead, the 'I' and the 'me' thus constitute an individual's personality. The interaction between a person's 'me' and her 'I' results in changes both to her personality and to the community with which she interacts, leading to new forms of social recognition. Mead expresses this as follows:

> Over against the 'me' is the 'I'. The individual not only has rights, but he has duties; he is not only a citizen, a member of the community, but he is one who reacts to this community and in his reactions to it, as we have seen in the conversations of gestures, changes it. The 'I' is the response of the individual to the attitude of the community as this appears in his own experience. His response to that organized attitude in turn changes it.
>
> (Mead 1934, 196)

In the process of mutual recognition, a person is being granted dignity and rights as a member of a social community (Honneth 1995, 79). However, a person is not only recognised as a member of the community, but she is also recognised as "a biographically individuated subject" (Honneth 1995, 80). Honneth adds: "The 'me' embodies the conventional norms that one must constantly try to expand, in order to give social expression to the impulsiveness and creativity of one's 'I'" (Honneth 1995, 82).

For the understanding of sustainable action in social contexts, it is particularly relevant that self-realisation should be understood as a process in which one develops abilities and traits. One can convince oneself of their unique value for the surrounding social world (e.g. the organisation, society) on the basis of the recognising reactions of other agents in interaction that are embodied in what Honneth calls "the generalized other" as part of the 'me'. Hence, it is a necessary condition for individual self-realisation that one is able to understand oneself as a unique and irreplaceable person. One cannot conceive of oneself as a unique and irreplaceable person until one's own manner of self-realisation is recognised by one's interaction partners to be a positive contribution to the community (Honneth 1995, 89). The struggle for recognition, which is what Honneth's account is about, could be viewed as the struggle between the agent and her environment to become a unique and irreplaceable person, who positively contributes to the community. Within the agent's personality, this struggle is reproduced between the 'I' and the 'me', where the 'I' represents the source of creative impulses, and the 'me' the generalised other.

The case studies in Chapter 3 are examples of moral actions that require a high level of social coordination (the semi-vegetarian in a one-person household also has to make arrangements with significant others). The efficaciousness of the agent depends on the willingness of other agents to cooperate. This means, importantly, that the agent must take into account the unpredictable reaction by other agents. That is, for morally relevant actions that are taking place in the public sphere, agents must creatively accommodate themselves to the reactions they encounter from other agents, in particular reactions by agents who are affected by the action under consideration. It is a general aspect of morally complex actions that the agent becomes aware of the psychical perhaps as early as the commitment phase, but certainly in the course of the executive phase. Consciousness of one's own subjectivity seems essential for agents involved in morally complicated situations, whether they aim for sustainable ends or not. Therefore it seems justified, when considering the case material, to look for differences as regards what happens in the agent's personality.

The struggle of an agent for her recognition in a social institution, which is mirrored in the struggle between the agent's 'I' and 'me', can become hard to bear, as a result of an increased discrepancy between the creative impulses of the 'I' and what is supported by the 'me' in its capacity as the generalised other. If the agent's deliberations and intentions that originate from the 'I' cannot sufficiently be reconciled with what is endorsed by her fellow agents, as far as this endorsement is represented by the 'me', there is a risk that the resultant psychological tension will become unbearable for the agent. Assuming that we have a situation of an agent in relation to many other agents, the only way for the agent to resolve the tension between her 'I' and 'me' is to withdraw entirely from the action situation. In that case, the struggle for recognition has become more like a battle the 'I' is losing. That is, the agent gives up her intention, ends her agency, stops being an agent in that particular social context.

We will review the case materials with regard to the risk that the psychological tension between an agent's 'I' and 'me' becomes unbearable, as a result of which the agent is forced to withdraw from the situation she has entered into by initiating her action. We will use the same ancillary concept of risk as defined in §5.2. That is, when considering the scenario of the agent being harmed by the tension between 'I' and 'me' to such an extent that it becomes unbearable for her, the risk that this will happen is defined as high when both severity and probability are high, or when one of them is high and the other is medium. The risk is considered medium when both severity and probability are medium, or when one of them is high and the other is low. It is considered low in all other cases. As before, probability is understood as the likelihood that the psychological tension between 'I' and 'me' becomes unbearable for the agent. Severity is understood as the impact on the agent's well-being of an increased tension between 'I' and 'me'. As seems natural to assume, an agent can bear a small discrepancy between the 'I' and the 'me', whereas a larger discrepancy will be increasingly difficult for her to endure.

As explained, both probability and severity are given the degrees low, medium, and high. We need to clarify the meaning of these degrees in the case of severity as follows. The impact that a tension between the 'I' and the 'me' has on an agent depends on how her action X and her end E relate to the social context in which she operates. If the agent can easily exchange X for an action Y for which no significant tension between 'I' and 'me' arises, whilst still achieving E within the same social context, a tension between 'I' and 'me' (with regard to X) will have a low severity. If the agent can resolve a tension between 'I' and 'me' arising for her end E by choosing to pursue, within the same social context, a different end F that fits within her overall ideal of a good life, a medium severity will be assigned. For example, on such grounds the (would-be) semi-vegetarian could decide to switch to an electric car as her contribution to mitigating climate change, instead of reducing her meat consumption. If the only way for the agent to reduce the tension between 'I' and 'me' is to step out of her social context entirely and start somewhere else to pursue E (or F), a high severity will be assigned to the tension between the 'I' and the 'me' in that action situation.

Turning to the case studies, the situation for agents in individual roles looks diverse. For the semi-vegetarian (see §3.2), the risk of an unbearable tension between the 'I' and the 'me' depends to a large extent on the composition of her household. Assuming she runs a one-person household, it seems that she is able to find a mode of realising the end she has set as regards, for example, social happenings where there is no vegetarian option available. If there is a partner or other household members, pursuing her goal could become more difficult. However, assuming that social relations are good, it seems unlikely that these would be harmed as a result only of the agent changing her diet. Ultimately, if no solution can be worked out, the semi-vegetarian could try to find another project as her contribution to mitigating climate change. For example, she could decide to concentrate on commuting by public transport, buying an electric car, or refraining from air travel over long distances. I deem both probability and severity of an unbearable tension between 'I' and 'me' to be low to medium. Consequently, the risk is also low to medium.

We have seen the impact on the agent's self-respect should he not be up to the job of being a doctor without borders (see §3.3, §5.5). So, assuming that the doctor is able to perform his role, we observe in the case material that this seems highly rewarding for him. He states: "I feel good about what I am doing. I want to do it better. Fixing other people seems to fix yourself. So, that is how I fix myself. That's what I found out." This indicates that the doctor integrated his action to join MSF into his ideal of a good life. Only if the doctor can no longer combine his commitment to MSF with his other obligations, for example with looking after his family, might he be forced to withdraw as a result of a tension between 'I' and 'me' that becomes unbearable. If that were to happen, donating money to MSF could also be an important way of helping to realise the mission of that institution, and could thus contribute to the doctor's ideal of a good life. Therefore, for the doctor I consider the risk of a psychological breakdown as a result of not being able to reconcile the 'I' with the 'me' to be low.

The "power grid rebels" (see §3.4) have embarked on a route towards a town that is free of nuclear energy. They face severe resistance from other inhabitants and from the regional power company. This means that the probability that their project will fail is at least medium. At the same time, the "power grid rebels" cannot easily exchange their project for another one that brings their end – stopping the dependency of their town on nuclear energy – in any way closer. Likewise, should they fail with the project, the only way for them to realise their goal would be to move to another town and start all over again. Hence, the severity of a tension between 'I' and 'me' seems high. This implies that the risk of tensions between the 'I' and the 'me' for these agents is high.

Although the credit restructuring officer (see §3.6) will many times run into morally dilemmatic situations, where the agent has a medium probability of psychological difficulties as a result of not being able to reconcile the 'I' and the 'me' with each other, there will also be alternative courses of action available. Alternatively, the situation and the rules and regulations at hand will lead to one or more alternative action situations. In addition, the agent will be able to find support for her course of action, as it is in the common interest of many other agents around her to minimise losses for the bank while maintaining the bank's reputation. As a result, I conclude that the risk of tensions between 'I' and 'me' will be low.

The situation of the purchase manager (see §3.5), however, is quite different. She has set herself the goal of reducing the company's dependency on non-renewable resources, with a more or less 'whatever it takes' attitude. At the same time, she has decided to initiate her action plan whilst carrying out her role as purchase manager. First, she encounters clear resistance from people who are more senior, and who could have the power to expel her from the company. The probability that this situation will lead to a tension between 'I' and 'me' for the purchase manager, therefore, is high. Second, as it is her role to be the purchase manager for oil and fat for the dairy foods company, she cannot easily exchange the project of changing to sustainable palm oil for another project that in a similarly meaningful way contributes to sustainability. This means that the only way for the purchase manager to reduce a potential tension between 'I' and 'me' is to step out of the action situation entirely – that is, by applying for a job with a company that agrees with her about the policy that palm oil must be produced sustainably in the first place. In conclusion, the risk of an unbearable situation for the purchase manager as a result of a tension between 'I' and 'me' is the highest of all the case studies we consider in this book.

## 6.6 Overview and diagnosis

As I noted for the commitment phase (see §5.6), the paradigmatic cases of sustainable actions in the executive phase are again the "power grid rebels" (see §3.4) and the purchase manager in the global dairy products company (see §3.5). The examples of these agents highlight most clearly the specific motivational hurdles that agents encounter when they are acting towards meeting Humanity's Challenge. These agents can only exercise a marginal influence on whether they succeed; they face a complex environment that does not allow for exercising unilateral effectiveness, irrespective of the competencies of the agent.

When we considered the risk of unreflectively changing one's intentions, I diagnosed that this risk is high for the paradigmatic agents compared with the other agents (i.e. the "power grid rebels" and the purchase manager vs other agents). These agents could form intentions and plans, and then refrain from executing, without anyone being affected. In order to withstand the resistance they encounter from other agents, they need to stand firmly for their intention by having a robust plan.

I continued to reflect on the complex social interactions that occur in the executive phase between the agent and other agents on whom she depends for her efficaciousness. These interactions potentially lead to motivational issues for the agent, arising from expectations by other agents to which the initiating agent does not live up. Other agents may withdraw their support following executive failures on behalf of the initiating agent; the initiating agent may suffer from a lack of self-efficacy or self-control; or the initiating agent may give up as a result of recognitional patterns that influence her interaction with the other agents. As examples of executive failure that relate to the agent's competencies, we discussed weakness of will and procrastination as phenomena that particularly affect the agent's effectiveness during the executive phase. Regarding weakness of will, we observed for the agents taking sustainable action that they have to reaffirm their intention repeatedly. These agents operate in social environments that are hostile to their plans. Weakness of will, as these agents display it, will not only be tolerated but may even be welcomed, as it is in the interests of the resisting agents that the action be delayed or cancelled. The paradigmatic agents seem more likely to display weakness of will and procrastination than the doctor without borders and the credit restructuring officer.

As to the sources of their self-efficacy belief, I observed that the paradigmatic agents rely only on good physical and mental health for support. Since both agents are able to manage their tasks until completion, it seems necessary to explain which source of motivation is available to them that enables them to overcome the motivational hurdles we have investigated. The self-control these agents perform seems to take the form of resisting reconsideration of their intention. The strength of their precommitment is revealed in the executive phase in that it enables them to resist the temptations to which they are prone. No other person or institution holds these agents accountable for not carrying out their sustainable action. This shows that agents who perform the paradigmatic sustainable actions have to rely on another source for their self-efficacy belief and self-control than those that have been discussed, which allows them to overcome the motivational challenges they encounter.

In Chapter 7, I will discuss a kind of commitment that may be at stake here. In my view, the paradigmatic agents make a kind of commitment to take up Humanity's Challenge that functions as a source of their self-control and their self-efficacy belief. I will propose that these agents make a so-called second-order commitment from which they receive an enhancement in self-control and a strengthening of their self-efficacy belief (see §7.4).

Finally, we observed that the struggle of an agent for her recognition in a social institution, which is mirrored in the struggle between the agent's 'I' and 'me', can become hard to endure. This seems to be due to an increased discrepancy between the creative impulses of the 'I' and what is supported by the 'me' in its capacity as the generalised other. If the agent's deliberations and intentions that originate from the 'I' cannot sufficiently be reconciled with what is endorsed by her fellow agents, then there is a risk that the psychological tension following from this lack of reconciliation becomes unbearable for the agent. Specifically, for the purchase manager this risk seems high in the sense that the psychological tension becomes so insupportable that the agent is forced to withdraw from the situation she has entered into by initiating her action. This could imply that the only way for the purchase manager to reduce a potential tension between 'I' and 'me' is to step out of the action situation entirely. That is, by applying for a job with a company that agrees with her about the policy that palm oil must be produced sustainably in the first place.

This overview and diagnosis of the executive phase concludes Part II. In Part III, we discuss what we can learn from this analysis. We will review, again, the perspective of the agent, and we will discuss examples of institutional interventions from the relevant literature that could benefit from the tentative hypothesis that can be developed on the basis of the conceptual-phenomenological analysis as carried out in Chapters 4 to 6.

## Notes

1 Vegetarianism (2016).
2 Korta and Perry call this "far-side pragmatics" (2015, §1).
3 Note that I will not enter into the debate about the question of whether akrasia, weakness of will, or procrastination can in some way be reduced to each other. See for a discussion Stroud (2010, 2014).
4 Note that Stroud formulates her definition on the basis of Holton (2009) and McIntyre (2006).
5 Holton defines 'strength of will' as follows:

> A person shows strength of will when they stick to their resolutions in circumstances in which they are right to do so; that is, when they do not reconsider them, and not doing so exhibits tendencies that it is reasonable for them to have. Strength of will turns to stubbornness when they stick by their resolutions even when it is reasonable to reconsider and revise them.
>
> (2009, 80)

6 The term 'psychical' is not Standard English. I borrow this archaic term from G.H. Mead.

# Part III

# Pathways for individuals, institutions, and humanity

# 7 Pathways for individual agents and institutions

## 7.1 Introduction

Of the four questions that characterise the problem with motivation for sustainable action, two have been crucial for developing the method of this book (see §1.3). The first question concerned the meta-ethical assumptions that are necessary for formulating the problem with motivation for sustainable action in the first place (see §1.4). The second question concerned the addressee of the normative requirement to take up Humanity's Challenge. To that question I responded that, in addition to the political institutions, individual agents in the affluent countries share *some* responsibility to take action in order to meet the challenge (see §2.2). From there, we proceeded with choosing the first-person perspective of the agent as the primary focus of our investigation (see §2.4). I argued that considering the agent's experience would help in proposing institutional policies that promote sustainable action.

The motivational challenges we have investigated by performing a conceptual-phenomenological analysis of the phases of the Rubicon model (i.e. in Part II) have twofold relevance. First, they form part of the practical self-understanding of people who take a sustainable action as *individual* agents, and who want to do something themselves in anticipation of the obstacles. Second, they are relevant for those who design and implement sustainable policies as *institutional* agents, since they need to take into account the perspectives of both those approving their policies and those affected by their policies.

Consequently, I want to identify capabilities for individual agents that allow them to cope with situations in which their motivation to proceed with their sustainable action is hampered. For example, we diagnosed that the paradigmatic agents in the commitment phase run into situations where their self-respect and self-esteem are challenged. Therefore, in order for these agents to be effective, their capacity for social coordination and recognition needs to be strongly developed. The pathway for individual agents will thus consist in a sketch of the practical self-understanding required for effective sustainable action. The autonomous agent in constitutional democracies then must decide whether she judges herself already sufficiently capable, whether she should start acting sustainably and develop the capabilities along the way ("learning by doing"), or whether she needs to develop the relevant capabilities prior to initiating her action (§7.2).

The agential capabilities form an essential part of what institutions need to know when they aim to steer people's behaviour towards taking up Humanity's Challenge. Following up on the need to support individual agents in acquiring the relevant capabilities, I will formulate institutional pathways for policies that promote sustainable action. In particular, it seems to me that moral education will become increasingly important for overcoming some of the motivational challenges for sustainable action. In addition, I will comment on informational and nudging strategies as institutional policies to steer agents towards sustainable behaviour (§7.3).

However, even when the relevant capabilities are in place and the relevant institutional policies implemented, taking sustainable action remains difficult. Holding on to one's commitment under contingent circumstances seems particularly hard to do in a context with multiple stakeholders. Agents in the affluent countries potentially remain ineffective, due to their seeming inability to deal with conflicting stakeholder interests. I will explain the extra commitment that is needed so as to overcome situations of inaction as a result of conflicting stakeholder interests, in terms of a second-order commitment. The main resolution agents need to take is that they *want to want to* take sustainable action. This entails our wanting to overcome the obstacles of which we have become aware by analysing our own efforts while taking sustainable action (§7.4).

## 7.2 Pathways for the individual agent

By going through the phases of the Rubicon model, we performed a conceptual-phenomenological analysis of the agent's perspective on her sustainable or otherwise morally complex actions. In this section, we will first group the elements of the descriptive diagnosis in one overview. In order to be able to start to formulate the capabilities for individual agents later on, we need such an overview. With the overview in hand, we will reconstruct, from the perspective of the individual agent, which capabilities are required to overcome the motivational challenges, whilst leaving the Rubicon model aside. Pathways for the individual agent entail that, if she is committed to take sustainable action, she needs to develop certain capabilities that enable her to deal with the motivational hurdles. These capabilities, then, can become the basis for proposing institutional policies that steer our behaviour towards acting sustainably.

A summary of the descriptive diagnosis forms the basis of our identification of the capabilities for individual agents. The *deliberative* phase is concerned with reaching an all-things-considered judgement from a number of prima facie judgements about available (sustainable as well as non-sustainable) actions. This is what the deliberation about the minor premises in a practical syllogism consists in. During this practical deliberation, the agent's instrumental rationality may be impaired, following errors of judgement, which are in turn caused by inadequate interpretations, for example of economic theories about the applicability of discounting. Such deliberative issues may lead to rejection of a given sustainable action proposal, or to indecision, both of which block the agent from moving to the commitment phase. There seem to be many ways in which agents could fail in being instrumentally rational. That is, there are potentially many cognitive obstacles that can hamper agents in reaching an all-things-considered judgement in favour of a sustainable action.

During their deliberations, agents are faced with two philosophical theories that facilitate them in being self-indulgent in the way they follow up on the major premise of the PSSA, which they have adopted (see §4.2). That is, agents ought to resist the temptation to be lenient with themselves, as facilitated by the point of view of common-sense morality that agents tend to take as their default moral position, and by certain conceptions of the human being with which we consider future people. Common-sense morality facilitates agents in having difficulties with vividly imagining future people as potential victims of their current non-sustainable behaviour. Moreover, it seems particularly difficult to replace the agent's experience of her non-sustainable behaviour with the experience of taking a sustainable action. The question about which conceptions of the human being we ought to adopt if we want to take seriously our efforts to meet Humanity's Challenge requires us to reconsider conceptions such as the Reductionist View. If the agent adheres to the Reductionist View of human nature, she is facilitated to believe that it does not matter whether she acts sustainably now, since she could delegate her decision to a future self. The Reductionist View has implications for legal and ethical concepts like desert, commitment, responsibility, and rights that we cannot accept. Instead, the conception of agency we need for working on Humanity's Challenge must allow incorporation of our actions and experiences as ours into the self-told story of our life over a long period of time.

From analysing the *commitment* phase, I take it that the cases of the "power grid rebels" (see §3.4) and the purchase manager in the dairy products company (see §3.5) are prime examples of sustainable actions with motivational challenges. First, the examples of these agents highlight the specific motivational difficulties that agents encounter when they intend to take action to meet Humanity's Challenge: they cannot relax on planning; they have to show means-end coherence all the time; and they are highly vulnerable to the belief that they will not succeed in their efforts (intention-belief incompleteness). Second, for these agents there seems to be a relationship between their commitment and their recognition by other agents with whom they socially interact. These agents have to cope with an at least temporary lack of recognition by other agents and a corresponding change in moral status. To set oneself on a course of sustainable action means that one needs the psychological resources of self-respect and self-esteem to withstand the potential undermining of one's commitment as a consequence of the conduct of other agents in reaction to one's commitment.

The diagnosis of the commitment phase has consequences for the formation of the agent's commitment as the existential moment of binding her will to follow practical reason and continue with executing the proposed action. For their sustainable actions, we conclude that these agents need to rely on their own strong commitment. We thus observe that agents taking a sustainable action need to rely on a strongly developed competency to bind their will to act according to their plan.

The analysis of the *executive* phase reveals that agents taking sustainable action act on the incentive of contributing to a sustainable world, which outweighs a number of incentives that do not favour such an action. One's choice to act upon the incentive in favour of a sustainable action constitutes one's practical identity in a different way than one's choice to follow incentives that do not favour sustainable action. When discussing weakness of will and procrastination we observe that agents taking sustainable action have to reaffirm their intention repeatedly.

The paradigmatic agents operate in social environments that in part are hostile to their projects. When considering the risk of nonreflectively changing one's intentions, our diagnosis is that this risk is high for the paradigmatic agents compared with the other agents. In order to withstand the resistance they encounter from other agents, they need to stand firmly for their intention by having a robust plan. In addition, we observe that the paradigmatic agents do not seem to have other sources to support their self-efficacy belief than good physical and mental health. The self-control these agents perform consists in their being able to resist reconsidering their intention. They have made a precommitment that has become a habit. It enables them to resist temptations of the relevant kind. As an explanation for the phenomenon of a precommitted agent who manages to deal with the motivational obstacles to sustainable action, we conjecture that these agents rely on another source for their self-control and self-efficacy belief than their physical and mental health only.

Still in the executive phase, we observe that the struggle of an agent for her recognition in a social institution can become hard to bear. Understanding such situations with the help of concepts from Honneth's theory of recognition, this seems to result from an increased discrepancy between the creative impulses of the 'I' and what is supported by the 'me' in its capacity as the generalised other. If the agent's deliberations and intentions that originate from the 'I' cannot sufficiently be reconciled with what is endorsed by her fellow agents, then there is a risk that the psychological tension following from this lack of reconciliation will become unbearable for the agent. This risk seems particularly high for the purchase manager, in the sense that the psychological tension becomes so unbearable that the agent is forced to withdraw from the situation she has entered into by initiating her action.

Knowing about the obstacles that may hamper our motivation means that we can take action to anticipate or prevent them from happening. We have formulated the problem with motivation for sustainable action for individual agents as "the problem of overcoming the inconsistency between the *moral claim* in favour of collective environmentally friendly action, and the *action* that we in fact carry out as an individual agent following up on that norm" (see §1.5). Thereafter, we investigated the inconsistency between the moral claim (i.e. Humanity's Challenge) and the action by performing the conceptual-phenomenological analysis of the agent's perspective. This analysis resulted in a descriptive diagnosis of the motivational problems agents are confronted with, which could explain the inconsistency mentioned in the problem formulation (assuming that they will be confirmed by empirical studies). Hence, the question we will now discuss is: How can individual agents do something about the motivational problems that they face when taking up Humanity's Challenge?

In my view, it seems that from the point of view of the individual agent this question concerns *capabilities* of coping with the motivational challenges for sustainable action. I thus understand the problem with motivation for sustainable action for individual agents as a problem of acquiring an appropriate set of capabilities – appropriate in the sense that these capabilities enable the agent to maintain her effectiveness, whilst dealing with the motivational hurdles. Without these capabilities, it may prove difficult for agents to solve the motivational issues and hence make overcoming them, to some extent, a matter of luck; obtaining these capabilities may at least provide agents with the real possibility of overcoming them through their own efforts. In other words, individual agents may need to improve their set of capabilities, if they commit themselves to act sustainably. Following the descriptive diagnosis, I cluster the agential capabilities for taking sustainable action into three groups.

First, agents need *capabilities of practical reasoning and judgement.* As we have seen, fully comprehending the deliberative complexities of problems concerning climate change seems particularly difficult (see §1.3, §4.2). There are numerous ways in which applying concepts and theories can push the agent's judgement-formation in such a way as to result in her choosing suboptimal or non-sustainable behaviour. Being instrumentally rational regarding possible actions that further the goal of meeting Humanity's Challenge and actions that do not thus seems a necessary capability for agents to reach appropriate judgements about action proposals. To mention a few examples: keeping one's knowledge of climate science up to date in order to remain confident about the scientific reasons in favour of sustainable action (see §1.3); being able to assess correctly the impact of one's actions on the climate, for example knowing that reducing one's air travel contributes significantly more to the reduction of climate change than changing one's lighting at home (Goodall 2011, 176); and being aware of one's psychological biases, for example when one's car provides one with a certain social status, instead of being primarily a means of transportation.

The agent's awareness that the common-sense phenomenology of agency may hamper her in choosing the best option for sustainable action, all things considered (see §4.3), seems an essential capability regarding practical reasoning and judgement-formation. Agents should be conscious of the background assumptions of common-sense morality, which may be appropriate in other cases, but often inappropriate when assessing options for sustainable action. I see two capabilities that may be required for agents to overcome the features of this phenomenology of agency. As agents tend to be less aware of the act of changing their behaviour in order to reduce their luxury emissions than they are of the act of continuing with these emissions, they require the capability of freeing themselves from this limited awareness of action options. For example, being able to view a bicycle ride as an option for acting that is as important as a drive in a sports utility vehicle. Another capability requires the agent to free herself from the spatial and temporal limits of her moral scope, thus focusing on the well-being, not only of certain significant people, but also of future generations.

Moreover, we must be aware of our perhaps hidden conceptions of the human being and the extent to which they play a role in our assessment and judgement about options for sustainable action. Such conceptions of the human being could cause us to step back in our reasoning, by eliciting sceptical worries about the major premise of taking up Humanity's Challenge. Thus, sustainable action requires the moral capability of resisting the temptation to follow along with lines of thought that justify agents in delegating their decision to a future self. As I have argued, this means that agents require the capability to resist the temptation to be self-indulgent as facilitated by the entertaining of two different conceptions of human nature: one for current people and a fundamentally different conception for future people (see §4.4).

Second, agents need *capabilities of social coordination and recognition*. As we have seen from the prime examples in the case studies of agents taking sustainable action, such agents need to be tolerant to the social frustration they may encounter in reaction to their initiative. These agents have to understand what their action may elicit in terms of reaction from significant other agents with whom they have to socially interact. The semi-vegetarian agent needs to be aware that when she attends a dinner with senior colleagues and managers, who may not underwrite her policy, these people may start to redefine their relationship with the agent by excluding the topic of sustainability from their commonly shared values. In that case, the agent needs to be capable of withstanding a perhaps temporary diminishing of her self-esteem (see §5.5).

Thus, agents should be able to understand and cope with the recognitional aspects of social cooperation, especially when a potential impairment of their self-respect and self-esteem is concerned. In addition, they need to understand and be able to deal with the risk of an unbearable situation for them as a result of a tension between the 'I' and the 'me'. Agents who set themselves on a course of sustainable action need the capability to withstand the potential impairment of their autonomy as a consequence of the conduct of other agents in reaction to their commitment.

Third, agents need *capabilities of the will.* The conceptual-phenomenological analysis reveals that agents taking a sustainable action require a will that enables them to proceed with acting on their intention to meet Humanities Challenge under circumstances that may be adverse. Under this cluster of agential capabilities, I subsume capabilities related to the forming of intentions, the making of plans, the mastering of instrumental coherence, the prevention of nonreflective reconsideration of one's intentions, being capable of strength of will, and avoiding procrastination. As we have seen, in the commitment phase, the paradigmatic agents taking sustainable action remain true to their commitment; they carry out their plans, they take the means to their ends, and they believe that they will follow through. In the executive phase, these agents do not seem to suffer from weakness of will or procrastination, and they do not nonreflectively reconsider their intentions such that this hampers them from realising their ends. Just as with the other clusters of capabilities, it seems crucial that agents who take up Humanity's Challenge possess a will with the above-mentioned features.

The pathway for the individual agent thus entails that she needs three groups of capabilities, with specific characteristics for the taking of sustainable action. If the agent is committed to taking up Humanity's Challenge, she will need to develop these capabilities with respect to the features that are required.

Now, one may ask: Who is responsible for the development of these capabilities? That is, given that we have accepted the collective duty to meet Humanity's Challenge, for which individual agents need the capabilities, then to what extent ought these agents to develop their own capacities, and to what extent should they be assisted? Again, just as individual agents in the affluent countries share *some* responsibility for taking sustainable action (see §2.2), I argue that these agents bear *some* responsibility for developing the capabilities required. As the agents have accepted the norm that sustainable action is required to mitigate climate change, and since we now have a view on which capabilities agents need in order to be effective, it follows from their autonomy that they bear some responsibility for developing the sketched capabilities. Their awareness of the need to acquire these capabilities makes them responsible for doing what they reasonably can in order to do so.

At the very least, if individual agents believe that they are not able, for whatever reason, to begin to develop the capabilities they need, they have a motive for being interested in institutional approaches that can provide assistance in developing these capabilities. Acknowledging one's need as individual agent to improve certain capabilities for taking sustainable action elicits one's interest in an institutional policy that aims precisely at strengthening these agential capabilities. Therefore, I will now discuss some roles that institutions can take up in the development of agential capabilities for taking sustainable action.

## 7.3 Pathways for institutions

The methodological premise of this book is that institutions that want to propose policies aimed at improving the conditions for sustainable action must make the point of view of individual agents their point of departure. Institutions that function within a democratic governance must start with an adequate understanding of the problems with motivation for sustainable action encountered by the individuals that are within their operative scope (see §2.4). These motivational problems have been investigated in Chapters 4 to 6, by analysing the phases of the Rubicon model.

Following on from that investigation, I have identified, independently from the Rubicon model, agential capabilities for dealing with the motivational hurdles (see §7.2). This means that different agential capabilities can be exercised in different phases of the Rubicon model. For example, the capabilities of practical reasoning and judgement will be required in the deliberative phase *and* in the other phases. Likewise, the capabilities of social coordination and recognition and the capabilities of the will are required in the commitment phase *and* in the executive phase. In addition, the motivational challenges do not apply to all agents with the same severity. For example, some agents will have a deficiency regarding the capabilities of practical reasoning and judgement, whereas others will need support to develop their capabilities of social coordination and recognition.

Therefore, institutional agents in their role as sustainable policymakers need to cope with a twofold complexity: first, at any moment in time, different agents are engaged in different Rubicon phases of taking sustainable action; second, different agents need different forms of support in order to overcome the motivational hurdles. Accordingly, as regards support for acquiring the agential capabilities, sustainable institutional policies will need to be relevant for agents in different Rubicon phases and with different needs. In my view, this means that the sustainable policymaker working on behalf of institutions such as governments, schools, and businesses will need to assist agents with various needs for motivational support in navigating through various Rubicon phases by implementing a *blend of policies* that facilitate sustainable behaviour. As a consequence, when discussing possible institutional interventions, I will not attempt to establish a relationship between the policy and the Rubicon phase for which the policy would be most appropriate. In this section, I will provide examples of how the institutional policymaker can make use of the investigation in this book by making the connection with philosophical and psychological research that has been done in the field of institutional policies.

As regards strategies that aim to encourage sustainable behaviour, the environmental psychologists Linda Steg and Charles Vlek distinguish between two categories of interventions. The first category, called informational strategies, is aimed at "changing perceptions, motivations, knowledge, and norms without actually changing the external context in which choices are made".

The second category consists of structural strategies. Here, changes in the circumstances under which behavioural choices are made "may be needed so as to increase individual opportunities to act pro-environmentally and to make pro-environmental choices relatively more attractive" (Steg and Vlek 2009, 313). Following this distinction, I will discuss how the tentative findings of the conceptual-phenomenological analysis in Part II can contribute to developing concrete intervention proposals.

First, I will discuss five informational strategies. These strategies are based on environmental psychological research. I will comment on these interventions and consider what contribution they make to enhancing the capabilities of individual agents that I have proposed as being necessary. It will become clear that these interventions are limited in their effectiveness.

Second, it is widely accepted that the informational interventions are especially effective when sustainable action is relatively convenient and the cost to the agent is low (ibid.). Therefore, structural institutional policies may be required that steer our behaviour towards sustainable behaviour. I will discuss a proposal by Ludger Heidbrink (2015), who discusses the advantages and disadvantages of institutional policies that aim at nudging an agent's behaviour towards more sustainable behaviour.

Third, even with the institutional policies discussed so far, working on meeting Humanity's Challenge will potentially result in insufficient progress. Recent developments in greenhouse gas emissions do not give reason to be optimistic about our current efforts to mitigate climate change (see §1.5). Therefore, I propose an institutional policy that aims at enhancing the agential capabilities by means of the moral raising and education of the younger generation. People need to learn the capabilities for effective sustainable action, which means that agents have to be educated in some way.

Note that by discussing the above-mentioned examples of institutional interventions I do not claim that these constitute a comprehensive set of policies that would be sufficient to meet Humanity's Challenge. For instance, I do not discuss questions of widespread distrust in the political institutions, or of the political leadership that would be required to engage the 'silent majority' in any affluent country to change towards sustainable behaviour. In other words, the purpose of this section is to discuss examples of institutional policies that could be supported by the descriptive diagnosis of the motivational challenges as developed in this book.

The task of taking up Humanity's Challenge requires us to strengthen the capabilities that help agents to overcome the motivational problems of taking sustainable action. The goal of strengthening these capabilities is to ensure that agents judge, commit to, and execute their action such that they are effective in acting sustainably. That is, the capabilities provide the agent with motivating reasons to stay on course with their sustainable action – by navigating through the Rubicon phase – and not deviate from it. The capabilities form reasons that explain the agent's behaviour in following through with the motive that the challenge provides, which I deemed weak for an agent *without* the capabilities (see §1.4). Within the context of what Dieter Birnbacher calls "moral pragmatics", the agential capabilities for dealing with the motivational hurdles thus serve as a means to the end of meeting Humanity's Challenge (Birnbacher 2015, 153).[1] I will now discuss intervention proposals from environmental psychology and show how these institutional policies can support the development of the agential capabilities.

Wokje Abrahamse and Ellen Matthies focus on five concrete proposals for the informational strategies mentioned above (Abrahamse and Matthies 2013, 224). As a first strategy they mention *provision of information*. This intervention aims to overcome the knowledge deficit about climate change and sustainable action in order to change people's behaviour. The body of climate scientific knowledge has grown rapidly over recent decades (see §1.3). However, one might question how much of this knowledge has reached the public at large, allowing people to understand the consequences of their own behaviour in relation to global climate change. It thus seems plausible to pay due attention to communicating the findings of climate science to a wide audience. Another kind of information, which is of use to agents, is information about ways to pursue their sustainable action. For example, knowing about recipes for vegetarian menus seems a prerequisite for the semi-vegetarian. However, as the empirical studies show, having the required knowledge does not do much to enhance the agent's effectiveness (Abrahamse and Matthies 2013, 225). Providing information about sustainable action by social role models may be more effective in stimulating behaviour change. The use of social role models may indeed enhance the agent's motivation, as it overlaps with Bandura's concept of vicarious experiences as a source of an agent's self-efficacy belief (see §6.4).

A second informational strategy is *goal setting*. This means that the high-level goal of meeting Humanity's Challenge is broken down into sub-goals, which are still ambitious, but also realistic from the agent's point of view. Goals must be achievable within a period of time, which the agent can oversee. The proposal of this book to strengthen agents' capabilities related to planning, which is particularly relevant to taking sustainable action, is compatible with this empirical result. Making sub-plans helps agents to achieve means-end coherence and to maintain the belief that they will succeed with their efforts (intention-belief incompleteness). The authors argue that so-called implementation intentions can enhance the effect of goal setting. These are "if-then plans that spell out, when, where and how a set goal has to be put into action" (Schweiger Gallo and Gollwitzer 2007 cited in Abrahamse and Matthies 2013, 227). Differently put, if the high-level goal of meeting Humanity's Challenge has been broken down into sub-goals, the agent needs an implementation intention to start executing the sub-plans.

The third intervention with empirical underpinning tests the agent's *determination to execute*. Agents are asked to sign a pledge to change their behaviour, on the assumption that their determination affects behaviour change via the reduction of cognitive dissonance. Cognitive dissonance is understood as "the tension that arises when one's beliefs or attitudes do not align with one's behaviour" (Abrahamse and Matthies 2013, 227). For example, I promised to do A, but I am doing B. Abrahamse and Matthies argue that making this pledge (e.g. to use public transport instead of the car) leads to a behaviour change, not only once, but also in follow-up situations (e.g. three months after having made the pledge). This intervention supports the proposal of this book to strengthen the agent's capabilities of the will related to making commitments and holding on to one's intentions. As the analysis of the Rubicon action phases reveals, making a commitment is needed if one is to change one's behaviour.

The fourth intervention is *prompting*. Here, the agent is confronted with "a short written message or sign, which draws attention to a specific behaviour in a given situation" – for example: "Thank you for not littering". When using this intervention, one assumes that the agent intends to carry out the behaviour in question, "but lacks a cue in the situation where the behaviour is required" (Abrahamse and Matthies 2013, 228). This intervention seems particularly relevant to preventing agents from nonreflectively reconsidering their intentions when they are executing a sustainable action. However, prompting is mainly effective in situations involving relatively easy behaviours, such as switching off lights or recycling household waste.

Finally, Abrahamse and Matthies suggest providing agents with *feedback* about their behaviour – for example, the power utility providing information about one's energy use over past years. The importance of feedback lies in the possibility for the agent to link certain outcomes to her behaviour that is necessary for reaching these outcomes. Indeed, feedback about the agent's achievements seems essential to prevent intention-belief incompleteness, particularly in diffuse situations of acting sustainably. This kind of information helps the agent with her belief that she can follow through. When agency situations are complex, feedback appears to be an effective way to encourage behaviour change, particularly when it is applied in relation to goal setting (Abrahamse and Matthies 2013, 229).

The above examples of informational strategies that institutions can employ all contribute to enhancing some of the agential capabilities for taking sustainable action. However, their contribution remains limited to only some of the capabilities of practical reasoning and capabilities of the will. The hard challenges of practical reasoning and judgement (e.g. temptations to follow common-sense morality and apply different conceptions of the human being) are not addressed. These interventions, moreover, do not help the agent to enhance her capabilities of social coordination and recognition (see §7.2). At the very least, the informational interventions need to be supplemented with structural intervention strategies that steer people towards adopting a more environmentally friendly lifestyle (Abrahamse and Matthies 2013, 230).

As the descriptive diagnosis in this book has revealed, keeping up one's motivation to take sustainable action is full of risks that this motivation will be hampered. As a consequence, one's sustainable action may not be completed;

the agent gets stuck in one of the Rubicon phases, meaning that the action stays suboptimal for some reason. That is, although the agent intends to take up Humanity's Challenge, she could have chosen a more sustainable act from among those that were at her disposal at the moment of making her choice (see §1.2). The motivational challenges tell us that there is a high probability that Humanity's Challenge will not be met as a result of the *individual* agent's impaired motivation. *Institutional* agents, who are responsible for the contribution of their institution to meeting Humanity's Challenge, thus have reason to propose policies that steer towards sustainable behaviour and that are independent of the motivation of individuals. That is, institutions might propose policies that make use of psychologically informed nudging in the sense that they influence the agent's decisions by modest means, without intervening in their freedom and restricting their autonomy (Heidbrink 2015, 174).[2]

The moral urgency to take up Humanity's Challenge arouses the debate about, on the one hand, institutional policies that circumvent people's motivational pitfalls, and, on the other hand, those that are compatible with individual liberty. How much moral pressure on agents is morally permissible, and where does nudging, understood as implementing behavioural institutional policies, turn into "illiberal paternalism, or worse moral dictatorship?" (Birnbacher 2015, 167). In the context of climate change, Ludger Heidbrink has made the interesting proposal to apply nudging as a structural institutional intervention in order to facilitate agents in acting sustainably. Here, I follow Heidbrink's discussion of the methodological and practical problems that make it questionable whether it is possible to promote sustainable action on the basis of psychologically informed nudging.

A basic assumption of nudging as an institutional intervention is that agents are influenced in their decision-making by contextual factors whose effects they are not explicitly aware of. Due to these biases, agents tend to make choices that are not in line with their own interests. As I have analysed, there are many circumstances in decision-making about options for action that cause individuals to decide in favour of actions that are suboptimal or even non-sustainable (see §1.2). Therefore, institutions have reason to seek political legitimation for policies that make use of nudging in order to steer agents towards sustainable behaviour. In this regard, a policy counts as 'paternalistic' if it attempts to influence the agent's choices in a way that facilitates her in taking sustainable action. However, the key libertarian aspect is that the agent's free choice remains in place and that the agent is under no coercion whatsoever. Thus, paternalistic measures are allowable if agents "are given the opportunity of *opting out* without thereby imposing undue costs or burdens on them" (Heidbrink 2015, 176, emphasis Heidbrink).

As Thaler and Sunstein argue, many agents can be considered "choice architects" in that they influence other agents by "organizing the context in which people make decisions" (Thaler and Sunstein 2008, 3). In fact, people always act in a socio-culturally shaped environment; they are bound by rules and have subjected themselves to contracts, standards, and routines that more or less directly guide their everyday actions. On the normativity of paternalistic measures Heidbrink states that "paternalistic measures are necessary to address behavioural anomalies, reduce information and knowledge deficits, to overcome weakness of will and lack of self-control and thereby to achieve the desired betterment of actors" (Heidbrink 2015, 177). Given this, libertarian paternalistic policymakers design the action situation of agents in such a way that agents are not forced to make certain decisions but need to make active choices if they do not want to be "nudged" towards promoting their own welfare. In sum, psychologically informed nudging aims to prevent agents from falling into the motivational pitfalls by an appropriate design of the choice architecture that provides motivational support for behavioural change – without restricting the agent's freedom of choice (Heidbrink 2015, 178).

On the basis of the above understanding, Heidbrink proposes the structural intervention of nudging agents to change their behaviour towards sustainable behaviour. This, however, does not mean that we should promote nudging without reservation. Heidbrink discusses three criticisms of what Thaler and Sunstein have called "libertarian paternalism" (Thaler and Sunstein 2008, 5).

First, libertarian paternalism is criticised for incoherence. There are significant doubts as to whether the conception of rationality underlying libertarian paternalism is compatible with its own basic assumptions (Heidbrink 2015, 182). The idea of designing the agent's choice architecture is based on the assumption that an agent can be modelled as a rational, informed, and self-disciplined homo economicus (see §4.2). However, this assumption is contrary to the diagnosis that agents are subjected to rationality defects and behavioural abnormalities in their action decisions due to contextual influences that prevent the realisation of stable preferences (Heidbrink 2015, 179). In other words, how can the paternalistic institutional policymaker be more advanced in knowing the agent's individual preferences than the agent herself?

If libertarian paternalism is applied in order to help agents stay on track with their motivation for sustainable action, then in my view this objection does not pose a problem. The purpose of having analysed phenomenologically (i.e. from the agent's point of view) the difficulties of taking sustainable action is that institutional agents can make use of this knowledge as a basis for their sustainable policies (see §2.4). That is, having integrated the motivational difficulties into one's practical self-understanding, one is able to accept that institutions influence one's choice architecture with the goal of helping one to overcoming those difficulties. Following the agent's acceptance of her own motivational weaknesses, when following up on her motive to take up Humanity's Challenge, she will accept the institutional nudges that steer her towards sustainable behaviour. The conceptual-phenomenological analysis of this book takes away the first criticism of libertarian paternalism, since the institutional policymaker and the agent rely on the *same* knowledge of the first-person perspective on the motivational issues with sustainable action.

The second criticism of libertarian paternalism is that it may lack efficiency. Does nudging lead to the envisaged improvement of the agent's welfare? As Heidbrink states: "The behaviour control over the design of choice architectures also requires that the designers are able to estimate the long-term consequences and expected utility of their interventions better than those who benefit from them." Again, the paternalistic institutional policymaker must assume that she is better in predicting the long-term effect of her policies on the individual agent than the agent herself, thereby taking the role of a "benevolent dictator" (Heidbrink 2015, 180). However, it would seem that the policymaker would make the same miscalculations in assessing practical situations as the agent does. In addition, policies tend to have unintended consequences, which the policymaker cannot prevent. In effect, the effectiveness of the applied control remains contingent and arbitrary.

In reply, note that for sustainable action we are not taking a view on the agent's preferences, but we are aiming to work towards meeting Humanity's Challenge. The purpose of the institutional control measures is to achieve this goal. Hence, if the policies are effective in mitigating climate change, based on functioning feedback mechanisms, they should remain in place. If these policies no longer contribute, then they must be replaced. In our case, the institutional control measures are not directed at improving the agent's welfare; they are directed towards the commonly shared goal of meeting Humanity's Challenge. In particular, the design of the measures is based on the motivational challenges for taking sustainable action as revealed by the descriptive diagnosis. Thus, the efficiency of nudging in our case is not based on the interpretation of contextual situations that influence the agent's welfare, but on its contribution to reaching a goal, which is independent from the agent's preferences. Thus, nudging for sustainability can be efficient, if it is based on knowledge of the motivational challenges for individual agents.

Third, libertarian paternalism is criticised for being illegitimate. In fact, it is said that libertarian paternalism is based on political coercion and unjustified limitations of individual freedom. The concern is that politically legitimate steering of behaviour is replaced by strategic coercion mechanisms, which seemingly only leave citizens with a choice between subjective action alternatives, while in reality forcing them to do what is necessary from the viewpoint of institutional welfare production. As a result, the critics state that nudging methods are illegitimate, since they are based, not on the autonomy and participation of the affected people in political decision-making, but on benevolent and non-transparent planning by institutions (Heidbrink 2015, 181).

Of course, one can never rule out the possibility of strategic coercion by institutional agents when these agents propose a policy for approval. However, in the case of a measure that facilitates sustainable action, the institutional agent will be able to legitimise her policies by referring to the commonly shared knowledge of the motivational problems with sustainable action. Since the motivational challenges have been investigated from the first-person point of view, it will be possible for the affected agents to verify whether the measure to be implemented indeed contributes to remedying the problem. Obviously, many people are poor at understanding their own motivations. However, the knowledge of the motivational challenges for individual agents allows the institutional policymaker to explain to the affected agents why a particular policy is proposed. In this way, an intervention to implement sustainability policies that is based on libertarian paternalism does not need to be in conflict with the principles of autonomy and political self-determination.

In discussing structural institutional interventions, I have argued that the project of taking up Humanity's Challenge does not need to be hampered by the criticisms of psychologically informed nudging. As Heidbrink argues, it can benefit from institutional policies that are based on libertarian paternalism. In fact, the descriptive diagnosis of the motivational problems with taking sustainable action in this book provides a foundation on which behavioural institutional policies to improve sustainable behaviour can be justified. If agents are confronted with their own motivational weaknesses when following up on their motive to take up Humanity's Challenge, they are more likely to accept the institutional nudges that steer towards sustainable behaviour. This is because the institutional policymaker and the agent rely on the *same* first-person knowledge of the motivational issues with sustainable action.

The institutional policymaker could argue that as a collective we have committed ourselves to take up Humanity's Challenge. However, through the conceptual-phenomenological analysis of this book we have come to know that as individual agents we are vulnerable to all sorts of motivational problems that hamper us in following up on our commitment. Therefore, as individual agents we have reason to support institutions that implement behavioural policies that prevent us from failing to live up to our commitment (see §7.2). The knowledge of our own motivational weaknesses gives us reason to propose sustainable self-binding by means of institutional policies that steer our individual behaviour towards the collective outcome that meets Humanity's Challenge. Here, self-binding should be understood as "the limitation of one's own freedom of action in order to achieve self-chosen aims in a better way" (Heidbrink 2015, 184). To conclude: using knowledge of the motivational challenges for individual agents, it is up to institutional agents to propose psychologically informed policies for sustainable behaviour that must be approved by the same individual agents in liberal democratic societies whose behaviour is the object of the policy.

The question remains, however, whether informational strategies and psychologically informed nudging policies indeed help to develop the agential *capabilities*, or whether they mainly correct behavioural anomalies. These interventions seem effective at least in enhancing some of the agential capabilities that I have proposed. However, they remain interventions that are imposed on the agent to the extent that there will be agents who cannot be reached by the informational strategies, or who are not supportive of the nudges for sustainable behaviour. These interventions influence the agent's behaviour by providing her with certain information, with one form of information having a greater effect than another. Or the choice architecture of the agent is influenced such that her behaviour is steered towards the preferred behaviour. It remains unclear, however, whether the agent's *motivation* to act sustainably is actually enhanced.

The main problem with interventions that are imposed is that the agent may fall back into her undesirable behaviour, once the intervention is interrupted or stopped for whatever reason (see §1.2). The risk of the interventions discussed so far is that they remain exogenous to the agent. They do not enable agents to *develop* their capabilities; they aim at correcting a chosen course of action, not at *building* the capability for making the choice in favour of sustainable action.

What seems to be lacking is a foundation, in the form of raising and educating people, for the agential capabilities to cope with the motivational challenges for sustainable action. The descriptive diagnosis makes clear that the capabilities are necessary (see §7.2), and agents may not possess them sufficiently. This means that I presuppose here that situations will have to be created in which these capabilities can be taught. In my view, this belongs to the institutional task of moral education. As Birnbacher phrases it:

> By educating the young generation in the spirit of sustainability and by creating an atmosphere in which cautious use of resources, nature conservation, and the long-term stability of social security are strengthened against countervailing short-term interests, society might build up resistance to its own tendencies to overuse resources and to infringe on the integrity of its own patrimony.
>
> (2015, 167)

Given the moral urgency of meeting Humanity's Challenge, I propose that the task of providing a pathway, in the form of moral education to enhance the capabilities, should be assigned to the relevant institutions.

In general, the task of moral education is understood as being to describe, to guide, and to reflect on moral development inside and outside school environments (Dietrich 2002, 423–24). Moral education entails value and virtue education, the formation of character and conscience, and education in ethical competencies. Moral education aims at facilitating the integration of the individual in society; however, it simultaneously aims at stimulating the dignity, autonomy, and responsibility of the individual. Given this, important questions in relation to moral education are: What can or must be taught and to whom? And, how and why ought this to happen? On the basis of the above understanding of moral education, I will discuss a proposal from the psychology of moral motivation, which shows how the development of the capabilities for sustainable action can be integrated into a curriculum of moral education.

In line with the concept of moral education, the psychologists Wolfgang Althof and Marvin Berkowitz have investigated which kind of education can contribute to developing the moral motivation of people. The authors propose a form of moral education that is supported by the Self-Determination Theory in psychology (Althof and Berkowitz 2013, 573).[3] This theory sees social and moral rules as initially external to the individual. In a slow developmental process, the individual more and more fully understands and internalises them. That is, over time the agent autonomously endorses the social and moral rules. Self-determined motivation, then, means that the agent approves her action as freely initiated and emanating from within the self (Althof and Berkowitz 2013, 570).

Althof and Berkowitz go on to emphasise that the motivation to act upon what one cognitively understands as the good is necessary to generate moral behaviour. Thus, doing the good requires a motivation to do the good, and any intervention, educational or otherwise, that is aimed at promoting moral development should include the promotion of moral motivation. This means, in their view, that students must develop a moral perspective, from which one understands the good. The moral perspective in part results from the experiences in school (and in other spheres of their lives) that they have integrated into their self-understanding and made compatible with other values and needs (Althof and Berkowitz 2013, 568). In accordance with the Self-Determination Theory, Althof and Berkowitz see the child as a self-defining organism with specific human needs that, when met, produce desirable motives that are central to the development of self (Althof and Berkowitz 2013, 573).

In summary, Althof and Berkowitz claim as the goal of moral education: "What we want as the result of schooling (and other socialisation of youth) are self-managed, self-aware individuals, who can and do act as competent moral agents. Especially in a democratic society, this is not only desirable but essential" (Althof and Berkowitz 2013, 574). In line with these authors, Christopher Schlottmann argues in favour of education *for* climate change, reflecting the interdisciplinary complexity and the ethical dimensions of climate change (Schlottmann 2014). So far, I have sought to explain why moral education in the agential capabilities is a necessary part of meeting Humanity's Challenge.

I will now discuss the main elements of the theory of Althof and Berkowitz about moral education and how their theory can potentially support education in the agential capabilities for sustainable action discussed above. Teaching the agential capabilities in schools means that schools must be organised intentionally to meet the need of developing these capabilities in children. The research by Althof and Berkowitz provides five conditions that serve as pedagogical advice for educating for moral motivation.

First, as explained above, the Self-Determination Theory assumes a gradual developmental process by which the person more and more fully understands and internalises the social and moral rules that were initially external to her. To enhance the capabilities of practical reasoning and judgement (see §7.2) it could be proposed that children be educated in the moral competency to judge what it means to strive for a sustainable future. This entails young children becoming competent to judge what it means to understand themselves within a chain of past and future generations. The concept of strong sustainability (see §1.2) could be taught as something to be valued for its own sake. Acting within the limits of the planet could be promoted as the norm. Later on, adolescents could learn to assume moral responsibility for their behaviour, whilst they are committed to work on meeting Humanity's Challenge. The teaching of these capabilities enables the agent in later phases of her life to deliberate about sustainable action among various other actions in such a way that misjudgements are prevented (Althof and Berkowitz 2013, 576).

An important feature of the capability of practical reasoning is that it extends the adolescent's moral horizon in order to avoid a limited awareness of the consequences of one's behaviour, and to open up her consciousness of the needs of future people. In the coming decades, children could learn about the transition of society towards meeting Humanity's Challenge. This means that an important question for moral education is: How can young people be taught to have a global moral perspective, which is required for working on Humanity's Challenge? (see §4.3). To enable young people to resist the temptation in later life to go for a ride in a sports utility vehicle just for fun, the climate footprint of the car drive is something that could be taught in schools. It seems important that children learn what it means to deliberate about whether to choose to go on a bicycle ride instead of a car drive. This could be done through providing examples of situations in which nature can be experienced and of how one can enjoy such experiences.

Another main topic in the moral education of agents in terms of the capabilities of practical reasoning and judgement is the influence of the transition towards meeting Humanity's Challenge on our conception of the human being. As I have argued, sustainable action requires the moral capability to treat the future bearers of certain rights according to the same conception of the human being as we treat the contemporary bearers of those rights (see §4.4). This means that students in schools could be made aware of the capability to resist the moral hypocrisy that causes them to entertain two different conceptions of human nature: one for current people and a fundamentally different conception for future people. I note that more research needs to be done regarding the what and how of this part of the moral education that is proposed. Unfortunately, this investigation must remain outside the scope of this book.

Second, Althof and Berkowitz argue that educating for pro-social relationships is essential to the development of moral motivation; and, third, they argue that moral motivation can best be promoted in pro-social communities (Althof and Berkowitz 2013, 577–578). These two conditions facilitate the development of the capabilities of social coordination and recognition (see §7.2). The capability of being tolerant to the social frustration that agents who take sustainable action may encounter can be built through mentoring in schools. Mentee and mentor regularly reflect on situations where the mentee does not feel understood in her social cooperation with others, specifically when sustainable action is concerned. Building an understanding of what a student's action may elicit by way of reaction from significant other agents could be done through education in communicative skills. This would lead to the ability to cope with the recognitional aspects of taking sustainable action, especially regarding a potential impairment of the student's self-respect and self-esteem. By involving students in school-wide activities with the goal of raising awareness of nature and the environment, the moral motivation for sustainable action could be enhanced. Thus, the school could take up the role of a community that embraces and practices the moral value of meeting Humanity's Challenge through promoting sustainable initiatives by its students.

Fourth, the education in moral motivation that could be provided by role models would seem to be most effective, according to Althof and Berkowitz's theory. Above, we discussed how providing information through social role models might be particularly effective in stimulating a change towards sustainable behaviour. Note that the use of social role models to enhance the agent's motivation is also supported by Bandura's theory of vicarious experiences as a source of an agent's self-efficacy belief (see §6.4). These theories together mean that, when focusing on the teaching of moral motivation for sustainable action, teachers in schools can stimulate the learning process of students by making the commitment to work on Humanity's Challenge and demonstrating the moral motives to do so. In fact, the development of all the groups of capabilities for sustainable action in children can be stimulated, if parents and teachers take up the role of being an example in walking the talk and talking the walk of sustainable action (Althof and Berkowitz 2013, 578).

Fifth, in the view of Althof and Berkowitz, in moral education it seems very important to educate for reflectivity and autonomous reasoning (Althof and Berkowitz 2013, 579). Through integrating a culture of reflection and autonomy, opportunities could be created for developing at least part of the capabilities of the will (see §7.2). These are capabilities related to the forming of intentions, the making of plans, mastering instrumental coherence, preventing nonreflective reconsideration of one's intentions, being capable of strength of will, and preventing oneself from procrastination. Allowing and even demanding a culture of reflection in schools could potentially contribute to strengthening the capabilities of the will in students, since their self-determined motivation, well-being, and engagement would be enhanced. However, more research needs to be done regarding the development of these capabilities in children, which falls outside the scope of this book.

Quite obviously, the purpose of the above discussion is to *start* the debate about what moral education can contribute to development of the agential capabilities for sustainable action. A lot more research needs to be done. For example, there seems to be a difference between what must be taught to very young children as opposed to adolescents and adults, and how this should be done. However, a discussion of these questions goes beyond what can be achieved in this book. At the same time, as I have shown, there seems to be a theory of moral education within the field of the psychology of moral motivation that could be of help in building the agential capabilities for sustainable action in young children and adolescents.

In this section we have discussed three kinds of institutional interventions that potentially help agents to carry out sustainable actions. These three interventions seem relevant to building up the capabilities in agents that are necessary to overcome the motivational hurdles of acting sustainably. As pointed out already, different agential capabilities may be exercised in different phases of the Rubicon model; and different agents need different forms of support in order to overcome the motivational hurdles. Therefore, the institutional sustainable policymaker needs to assist agents who have various needs for motivational support in navigating through the various Rubicon phases by implementing a blend of policies that support sustainable behaviour.

However important the institutional policies are, though, they cannot prevent situations of severe inaction, in which, in fact, none of the institutional parties can be blamed for this inaction. What must agents do when the problem of inaction becomes so structural that all the parties involved can blame 'the situation' for the lack of progress? In the next section, we try to reach an understanding of what an agent, who wants to take up Humanity's Challenge in an institutional context, and who has run into a situation of institutional inaction, needs. This will prepare us for Chapter 8, where we will elaborate a case study of an agent acting sustainably in an environment with multiple stakeholders.

## 7.4 The need for a second-order commitment

So far in this chapter, we have proposed agential capabilities that individual agents should be interested in developing with the goal of being effective in taking sustainable action while they proceed through the phases of the Rubicon model. Subsequently, we have proposed institutional policies that facilitate sustainable behaviour and the building of the capabilities in agents. Since the goal of both the agential capabilities and the institutional policies is to prevent the agent from being hampered by the motivational hurdles, they can be considered to be measures that take away what Cass Sunstein and Edna Ullmann-Margalit call "on-the-spot burdens" of taking sustainable action, by accepting some "ex ante burdens" (1999, 14). That is, developing the agential capabilities and implementing the institutional policies, seen as ex ante burdens, facilitate the agent in overcoming the motivational challenges (i.e. the on-the-spot burdens) of acting sustainably.

However, even if individual agents have done all that can reasonably be required to anticipate the motivational roadblocks, and even if institutional support is provided, another obstacle looms large that we can take away only by further strengthening our commitment to meet Humanity's Challenge. This happens when, despite the ex ante measures taken, the on-the-spot burdens become a lot heavier than the agent anticipated. Sunstein and Ullmann-Margalit take as an example the decision as to when to wage or terminate a war, which probably requires overcoming heavy ex ante burdens, whereas the on-the-spot challenges remain large as well (ibid.). The obstacle I have in mind in the context of meeting Humanity's Challenge is severe inaction or *stalemate*. That is, I am interested in what happens to the agent's motivation when her sustainable action brings her to a situation of severe inaction or stalemate, while being equipped with the agential capabilities and supported by the institutional policies.

Many agents taking sustainable and other moral actions experience problems with carrying out their plans. On the one hand, there are problems that the agent can solve relatively straightforwardly, when she has gained the required capabilities (see §7.2). For example, if the semi-vegetarian knows what to cook for dinner, she can buy the ingredients and start cooking, even if she is preparing this vegetarian dinner for the first time. On the other hand, there also seem to be problems of a different kind than that of trying to ensure one is instrumentally coherent in the sense of taking the means to one's ends or preventing oneself from procrastination. These are situations in which, even if all the parties have tried hard, through exercising the required capabilities as individual agents, while being supported by institutional policies, the institutional framework is just not suited to dealing effectively with Humanity's Challenge.

As an example of the latter sort of problems, in Chapter 8 we will discuss the case of an agent in the banking industry, who is engaged in relationships with multiple stakeholders whose diverse interests cannot be fulfilled immediately and straightforwardly. For this agent the question to be dealt with is: If the bank's sustainable policy leads to the need to redefine the stakeholder relationships, then what is required of her in order to prevent a situation of stalemate? In my view, situations such as the banker's require a *strategic or political* capability to resolve situations of stalemate with a more or less 'whatever it takes' approach. In order to reach a new equilibrium in the relationships with their stakeholders, these agents will need to show a commitment to proceed with implementing the sustainable policy, which may be stronger than the commitment we have seen to be required for sustainable action generally, which in any case we have supposed them to already have made.

As we have discussed (see §5.2), committing oneself to take action involves answering questions such as: Do I want to take this action? How will engaging in this action change my life? Knowing the simple fact that I alone can initiate my actions, am I prepared to do what is required of me in order to act efficaciously? The agent faces the moment of her resolve, assuming moral responsibility for what she is going to do. While making a commitment to act in accordance with Humanity's Challenge, the agent's moral responsibility is being called upon in a specific way: Am I prepared to do or refrain from doing certain things in order to meet the challenge, even though my actions will primarily benefit people who will live in the distant future? However, in situations in which actions must be taken that require strategic or political capabilities, it seems that the agent's commitment needs to be strong enough so as to face a further question: Am I prepared to take moral responsibility for the changes in the institutional set-up that are required to reach the new equilibrium in relationships with stakeholders, including the consequences this might have for other agents, even though my actions will primarily benefit people who will live in the distant future?

As I see it, Sunstein and Ullmann-Margalit come close to describing the form of such a commitment when they define what they call "second-order decisions":

> By 'second-order decisions' we refer to decisions about the appropriate strategy for reducing the problems associated with making a first-order decision. Second-order decisions thus include strategies that people use in order to avoid getting into an ordinary decision-making situation in the first instance.
>
> (Sunstein and Ullmann-Margalit 1999, 7)

That is, in addition to acquiring the agential capabilities and being supported by the institutional policies, the agent makes a particular meta-decision to overcome the motivational problems that she will come across on the spot when acting sustainably. As to the content of this commitment, along the lines of Harry Frankfurt's famous view, I argue that making the meta-decision as a second-order commitment to take sustainable action may enable agents to overcome the heavy on-the-spot motivational challenges. The strategic or political capability that comes to expression in agency situations that are hampered by severe inaction or stalemate, in my view, can be equated with making a second-order commitment to take sustainable action. Let me clarify this second-order commitment as follows.

After investigating the agent's first-person perspective, we have obtained a descriptive diagnosis of the problems with motivation for sustainable action (see Part II). We are now aware that our motivation for pursuing a sustainable course of action will be challenged; for each of the phases of the Rubicon model we also know some of the specific problems that might occur. In addition, we now have a picture of the sorts of capabilities that agents require to overcome these motivational problems. Importantly, this means that the agent's first-order commitment to sustainable action in the form of "A wants to X" may not be strong enough "to play a decisive role in what A actually does or tries to do" (Frankfurt 1971, 8). Note that Frankfurt talks here about "first-order desires". Since in "A wants to X" the term 'to X' refers to an action, in my view the statement "A wants to X" can also be understood as a first-order commitment of A to X. For an agent to develop the required capabilities for sustainable action on her own, and for resolving problems of severe inaction or stalemate in multiple stakeholder environments, a commitment is required in the form of "A wants to want to X" (Frankfurt 1971, 10), which Frankfurt calls a "second-order volition".

Thus, I understand making a second-order commitment to take up Humanity's Challenge as making a commitment to meeting the motivational difficulties that one will face, without losing the sensibilities to learn from one's mistakes and the ability to receive feedback from other agents. The agent does not only commit to take a sustainable action; she also commits to overcome the motivational difficulties of which she has become aware. The second-order commitment constitutes our attitude to overcoming the motivational hurdles, of which we have become aware through performing the conceptual-phenomenological analysis. Similar to how Frankfurt has formulated it, the second-order commitment means that agents want to want to take sustainable action. Such second-order commitment not only raises the question for a specific action (Do I want to take this action?), it goes further in that it determines the agent's will. The second-order commitment ensures that the agent does everything that is required to make sure that she does not stop (first-order) wanting X – and thus, even in the case of a stalemate, will not give up on this first-order commitment. Such second-order commitment may enable the agent to cope with the significant risks, which I elaborate in the case of the banker (Chapter 8), of which she perhaps was not aware when she committed to taking the action of promoting the sustainable policy.

At the very least, the second-order commitment to want to want to take sustainable action means that one refrains from taking non-sustainable actions without which one's life remains well above a minimum threshold of subsistence. For example, wasteful driving in a gas-guzzling SUV on a sunny Sunday afternoon (see §1.2) should be among the behaviours that we end when we have made the second-order commitment to help meeting Humanity's Challenge. In other words, when the second-order commitment is made, it will be rational of the agent to accept the on-the-spot burden of refraining from non-sustainable behaviour that does not contribute to one's ideal of a good life.

Our considerations make clear that the agent's commitment, as her existential choice to act sustainably, is crucial for making progress with taking up Humanity's Challenge. Despite the support that institutional policies can provide, our decision to engage in sustainable action cannot be enforced by anyone. We cannot outsource our commitment. The problem of mitigating climate change will not be solved unless we as agents in the affluent countries realise that it is up to us to make the commitment to take sustainable action.

However, this is not to say that the problem with an institutional set-up that keeps us from making progress with meeting Humanity's Challenge can be delegated to individual and institutional agents only. The success of such delegation would end up depending on the existential courage of these agents. As we will see in the case study of the bank manager (Chapter 8), an extra commitment is indeed needed for pursuing the sustainable policy; there may, however, be changes needed in the institutional infrastructure that go beyond what one can reasonably demand from an institutional role occupant. In this book, however, I stick to my choice of a minimally individualistic concept of institutions (see §2.2), and, consequently, I leave a further discussion about institutional requirements for meeting Humanity's Challenge aside.

If the commitment to act sustainably forms the step that agents cannot leave to anyone else, then we can consider the situation in which agents fail to make this step. The lack of progress in mitigating climate change could be the result of our failure as individual agents to make the commitment to take up Humanity's Challenge. I will consider this situation in Chapter 9. I will argue for a specific understanding of the concept of moral corruption that elaborates what it means *not* to make the commitment to take sustainable action. I then argue that we cannot strive for our ideal of a good life, which includes viewing ourselves as persons of moral integrity, if we adhere to a morally corrupt course of conduct.

## Notes

1 Birnbacher (2015, 153) understands moral pragmatics as "primarily concerned with questions concerning the means of implementing moral principles under real-world conditions".

2 Note that Heidbrink speaks about these policies against the background of "libertarian paternalism".

3 The authors contrast the Self-Determination Theory (SDT) with the Positive Behaviour Interventions and Supports (PBIS) approach. This theory emphasises extrinsic motivation by seeing the child as a set of behaviours in a context that either supports or thwarts those behaviours. Such set of behaviours is defined, a curriculum teaching the behaviours is implemented, and a clear set of consequences for misbehaviour and appropriate behaviour is applied. In their discussion about the choice between SDT and PBIS, the authors argue that intrinsic motivation has a stronger influence on an agent's behaviour. Therefore, Althof and Berkowitz choose in favour of SDT and, as a consequence, I go no further into discussing PBIS in this book.

# 8 A sustainable policy for the banking sector

## 8.1 Introduction

We have conceptualised the second-order commitment to take up Humanity's Challenge as that one makes a commitment to meeting the motivational difficulties that one will face (see §7.4). The agent does not only commit to taking sustainable action; she also commits to overcoming the motivational difficulties of which she has become aware. The second-order commitment ensures that the agent does everything that is required to make sure that she does not stop (first-order) wanting the sustainable action – and thus, even in the case of stalemate, will not give up on her first-order commitment. Given that such actions take place in social contexts, this means that the agent, although she displays her commitment, intends not to lose the sensibilities to learn from her mistakes and the ability to receive feedback from other agents.

In order to illustrate situations where second-order commitments arise, we will discuss in depth the example of an agent in the banking industry who faces the problem of severe inaction or stalemate because of stakeholder interests drifting apart. We ask: What are the ethical margins of manoeuvre for bankers, when implementing a policy for improving sustainable behaviour, given that retail clients, business clients, employees, regulators, and shareholders heavily influence them? As the case study will show, agents can become caught in a web of stakeholder interests, which requires them to make a second-order commitment in order to prevent the on-the-spot burden of severe inaction. At the same time, once the second-order commitment is made and the agent receives initial support from stakeholders, the case study shows the potential for implementing sustainable policies in the banking sector.[1]

The case study concerns an institutional agent – a bank manager – who, similar to the purchase manager in the dairy foods company (see §3.5), is committed to changing her institution's policy so as to contribute to working on Humanity's Challenge. I have chosen a case study from the banking industry since this industry can be characterised by institutional agents who operate within a well-defined set of existential stakeholder relationships. Furthermore, the agent's situation and the role descriptions of the stakeholders are applicable across the affluent countries. If we consider a bank as a nexus of contracts between a network of individual agents and parties that have an interest in the company (i.e. stakeholders), then what will be the consequences for these stakeholders should the bank adopt a policy that promotes sustainable banking? I will describe the agent in her role as bank manager within this network of stakeholder relationships – again from the agent's point of view. These stakeholder relationships are vital: the bank and the agent in her role as bank manager cannot do without them (§8.2).

With the case study, I want to make clear how agents can run into situations in which they feel an existential moment of uncertainty about how they should proceed with living up to their commitment to act sustainably. That is, the agent is aware that the institutional policies have, until this moment, supported her adequately, and that her agential capabilities have been sufficient to cope with the challenges she has faced so far. However, she is now confronted with a challenge she could not have anticipated, at a time when her psychological resources are depleted. In particular, my goal is to show how an agent can encounter a situation in which her individual capabilities are worn out, in which the institutional agents on whose cooperation she depends are justified in not cooperating any further, and in which she has no power whatsoever to enforce her effectiveness (§8.3).

After having described the agent's situation, which she can overcome only by exercising her second-order commitment, we look at a real-life example of a bank that has implemented a sustainable finance policy. The policy entails business clients being rewarded, by means of a lower interest rate on their loan, for realising their commitment to improve on sustainability. The events that followed the launch of the policy provide an example of how an institutionalised nudge (see §7.3) can elicit behavioural change in the banking sector – in this case in the form of other banks adopting the same sustainable finance policy and a banking industry body formulating a standard based on it. The impact of the bank's sustainable finance policy on the sustainable actions of corporations will be assessed and further challenges will be discussed (§8.4).

## 8.2 A bank as a nexus of legal and ethical relationships

According to classic economic theory, a firm can be seen as a nexus of contracts among the parties that constitute a corporation (Boatright 2008, 177; Jensen and Meckling 1976). Economic theory further assumes that markets function optimally when actors are motivated by self-interest without recourse to ethical motivation (Koslowski 2012, 4). We will not discuss here the extent to which the motive of self-interest is a driving force in other sectors of the economy. In the case of banks, however, I argue that, alongside the motive of self-interest, ethical features inevitably do form part of the nexus of contractual relationships that constitute the corporation.

If we put our money into a bank, it is obvious that we hand over something which is of value to us to a party that is outside of our direct control. We put the savings that we consider instrumental to realising our ideal of a good life into the care of an institution, which is legally entitled to use the funds for its purposes. As we make the deposit, it seems entirely reasonable of us to expect that the bank will be able to return the money at a later date, together with interest in return for the bank having used it in the meantime. We thus put an amount of trust into our contractual relationship with the institution we hand our savings over to. Scandals in the financial world shock us because they involve people and institutions in which we have a legitimate expectation of trust (Boatright 2008, 4). The involvement of trust in the relationship between a bank and its clients shows that this relationship cannot be defined entirely in legal terms. Ethical standards inevitably play a role. The relationship between a bank and those parties that have a stake in the bank's activities is affected by the moral integrity of individuals, the ethical culture of the institution, and the moral leadership of people in positions of responsibility.

If a bank decides to follow a policy of engaging in sustainable action, then such a policy must not only comply with the legal standards governing its relationship with its stakeholders; it must also fulfil the ethical standards that are implicit in this relationship (Boatright 2008, 27). In order to understand the motivational difficulties for bankers when they decide to implement a policy that promotes sustainable action, we need some insight into the ethical issues bankers have to cope with in dealing with their most prominent stakeholders. In this section, we consider retail clients, business clients, employees, and regulators. The last part of this section is devoted to the shareholders. Since this latter group can be characterised as the residual claimholders, meaning that their interests are served *after* the interests of all other stakeholders have been respected, a separate discussion about these stakeholders is warranted: how they influence the behaviour of bankers, and the duties that bankers have towards them.[2]

Moreover, I have assumed that the bank in this case study operates in markets under full competition. This means that the bank cannot take advantage of its position in the market. It cannot dictate prices or conditions; both retail and business clients have ample opportunities to change to another bank; employees can switch to another employer; regulators can withdraw the bank's licence to operate; and shareholders can buy and sell shares at any moment they wish.

*Retail clients* are usually individuals who use a bank for dealing with money-related matters in their daily lives. This typically involves the provision of payments, savings, credit, and investment services. Banks have the fiduciary duty to their clients to care for what is a source of value earned or inherited by these clients during their lives. What, then, are some of the legitimate expectations of retail customers that bankers have to live up to? In addition to expecting that the bank will return their savings money, with interest, retail clients may expect that the bank will refrain from servicing sectors of the economy whose activities are in some way morally impermissible. For example, is there a moral problem if the bank lends on the money from a trade union to a company that heavily uses child labour under questionable circumstances? Or, would it be morally problematic to lend on the money entrusted by a pacifist retail client to a producer of cluster bombs? How do these potential restrictions on the bank's options for lending relate to the client's demand for the return of her money together with interest that has been agreed under competitive conditions?

These questions bring to the fore the problem that there could be a discrepancy between the intentions of the retail client about what ought to be done with her money in a way that remains consistent with her ideal of a good life, and the opportunities for the bank to realise a return as expected by the same retail client. Given the principles these clients have adopted (i.e. rejecting child labour abuse, disapproving of the production of offensive weapons), it seems likely that they would judge it morally wrong should they learn that the bank was lending their savings money to companies that make use of child labour or that produce cluster bombs. At the same time, these clients might follow economic rationality in the sense that they would not hesitate to withdraw their money if they were to detect an unfavourable interest rate, compared to another bank.

This means that bank managers face at least two ethical challenges in their relationship with retail customers. First, they must lend the money entrusted to them in such a way that they are able at all times to pay the deposit holder back, including an interest margin. This, in turn, explains the attitude of bankers towards business clients, namely that loans (i.e. principal and interest) must be paid back on time. Banks are risk averse by nature, since they are aware that incurring losses on loans extended to business clients could endanger the bank's own existence. Second, the banker faces the restriction that the money entrusted to her cannot be lent to simply anybody who is able to pay interest and principal. Although banks are legally entitled to lend to many business clients, there are moral restrictions that prevent the bank from lending to a corporate customer that is involved in abuse of child labour or the production of cluster bombs. Towards its retail customers, the bank thus runs the risk of damaging its reputation as an institution that cares for its moral values, should customers find out that the bank is engaged in lending to businesses engaged in morally impermissible activities. However, given the economic paradigm of rational self-interest, for the bank it is not always clear which moral standard prevails in the eyes of the retail customer. Is it the interest rate and the immediate availability of the entrusted funds, or is it the socially responsible lending practice of the bank that is most important to them?

*Business clients* – that is, corporations, foundations, and (semi-)sovereign entities – are run by groups of people who need banks if they are to perform their role in the economic and governmental sphere. The relationship between bank and business client has several ethical features. First, before entering into a relationship with a company, the client needs to disclose information, which may be publicly available or not, about its ownership structure, and whether it is in compliance with relevant laws and regulations. Second, in the case of a business that applies for a loan, clients must provide detailed records about their financial situation. The bank's credit analysts are specialised in spotting strengths and weaknesses in the company's finances, and in the organisational structure of the corporation. Third, the company must disclose in detail the purpose for which the credit is needed. This means that the client's plans for future investments have to be disclosed to the bank. For example, these plans could entail taking over a competitor. In particular, the business case for the investment should become clear. In other words, the customer must show the bank its detailed plans for how the investment will be paid back, which sources of income will be used, what the operating costs will be, and how much cash flow will be generated.

In general, business clients need different – more complicated – financial instruments than the retail customer. For example, they need working capital facilities to ensure their short-term liquidity, and they need loans with longer tenors to finance the investments that they cannot finance out of their own cash flow. By means of attracting bank loans, corporations can leverage their equity, which potentially increases the value of the firm to the shareholders, if the business case of their investments is realised. In addition, business clients use banks as an intermediary to assist in obtaining debt or equity directly from investors (e.g. pension funds) who operate in the capital markets.

Importantly, bankers judge potential business clients on their integrity. Apart from the question whether the client will be *able* to pay back the loan in terms of available cash flow, the question is raised: Does the client *intend* to pay back the loan? Bankers are experienced in assessing whether the client has somehow expressed her intention to pay back the loan, as this is savings money that belongs to the retail clients. In credit decisions there is no trade-off between economic arguments and the integrity argument in order to finance a particular client. Both are necessary conditions that have to be fulfilled. The question whether the business client acts in good faith towards the bank plays a crucial role in credit decisions.

The bank faces similar moral uncertainty as it does with the retail client, when it comes to the business client's commitment to meet Humanity's Challenge. Suppose a business client approaches the bank with a request for finance for an investment that falls within the preferences of the bank's retail clients to support sustainable business. Does this imply that the loan must be offered at the most competitive interest rate as well? That is, although the client and the bank may agree about the purpose of the credit, this does not mean that the client should be happy with the commercial conditions offered by the bank. Thus, it will be instrumentally rational for the client to look for another bank that offers a lower interest rate, even though she was in agreement with the previous bank about the moral aspects of the credit. From the business client's perspective, moral reasons and pure economic reasons compete with each other for prevalence.

From the above considerations, and an assumption of full competition, it becomes clear that a bank policy that promotes sustainability has to be commercially attractive to business clients. Customer investments that are needed to mitigate climate change have to be financed at the lowest possible economic costs. The bank's policy for sustainability must be acceptable to the business client *from both a moral and an economic* point of view. If it only meets one of these requirements, it runs the risk of not being adopted at all; that is, given the weak motivational force of the moral motive to meet Humanity's Challenge (see §1.4), the economic motive of obtaining credit at the lowest possible costs is then likely to prevail over the moral motive to promote sustainable action.

*Employees* in a bank assume responsibility for maintaining the economic and ethical aspects of the relationships with retail and business customers. Towards the retail client, the banker is accountable for ensuring that the saved money generates a return that is in conformity with market practice, and that it is not invested in loans to companies that engage in morally impermissible activities. Towards business clients, the banker is accountable in two respects. First, a bank loan can be among the necessary conditions for the business continuity of the borrower. This means that the banker is accountable to the borrower directly, meaning that she must agree with the borrower the conditions under which the loan is extended. Second, the banker is accountable to the borrower's stakeholders in an indirect way: the bank enables the borrower's activities to take place and thereby takes responsibility for the consequences of those activities for the customer's stakeholders. Banks thus partake of the moral responsibility for the fabric of activities constituted by their borrowers and the parties those borrowers do business with.

There is, however, a further responsibility that bank employees have to live up to, which is ensuring the continuity of the bank itself. Not only are the bank's business clients under competitive pressure, but the bank itself operates in competitive markets. For customers, this means that they may be able to improve their situation by switching to another bank. Like any business operating in a competitive market, the bank has to satisfy the requirements of its clients, and, simultaneously, ensure that the bank realises a profit on the services it provides. This implies that bankers must close transactions that will provide them with an income from which to pay salaries, expenses, and the costs of borrowers that are not able to pay the debt service on their loans (i.e. risk costs). In particular, bankers have to ensure that their agency costs are compatible with market practice, in order to realise a return on their capital, as their shareholders require.

*Regulators* of banks and the financial markets represent the interest of the public at large in the continuity of the financial system. The public's interest must be serviced with financial products that are easy to understand and do not confront customers with unanticipated risks. The task of regulators is to ensure the continued existence of the financial infrastructure that enables society to function as an economy. In Europe, the European Central Bank (ECB) has become responsible for key tasks concerning the prudential supervision of banks. It has the power, among other things, to grant or withdraw the authorisation of banks in the euro area; assess acquisition and disposal of (share) holdings in banks; and ensure compliance with prudential requirements laid down in EU banking rules and set, where necessary, higher prudential requirements for banks to protect financial stability.

To embed the ECB in a democratic institutional framework that represents the public's interest, a number of accountability arrangements have been put in place to hold the ECB accountable for its supervisory tasks. These include regular reporting to the European Parliament, the EU Council, the European Commission, the Eurogroup, and the national parliaments of the participating member states; auditing by the European Court of Auditors; legal control by the Court of Justice of the EU.

Following the Paris Agreement (2015), the Dutch Central Bank (DNB) issued a study highlighting the consequences for the Dutch financial sector of a transition towards a carbon neutral economy. DNB emphasises that, providing the relevant institutions reach an agreement at short notice, a gradual transition to a low carbon society seems feasible. A green economy requires different sorts of capital goods. If the economic lifetime of carbon intensive industrial processes falls within the time horizon for meeting Humanity's Challenge, then the transition can be made at reasonable costs. However, if parties do not reach a commonly shared vision as to how to implement the climate goals, then an abrupt transition will be likely, causing a sudden depreciation of carbon intensive companies (DNB 2016, 46). Should this scenario happen, then banks will probably have to write off loans outstanding to these firms, which in turn will result in an increased probability of banks going into default.

DNB (2016, 47) calls this the scenario of "the carbon bubble hypothesis". The hypothesis states that, as a result of the ambitious goals for mitigating climate change as formulated in the Paris Agreement (UNFCCC 2015), the value of current fossil fuel reserves is overstated. An unknown part of these reserves will not be developed and, consequently, will no longer have economic value. However, it may still have a value in the books of some oil & gas companies. In a gradual transition scenario, the resulting losses from writing off these assets can be absorbed; in an abrupt transition scenario, however, an accelerated depreciation increases the probability of corporate default.

In summary, bankers are in a continuous dialogue with their regulators, who provide bankers with their licence to operate. As with the other stakeholders, there is no way that bank managers can act without approval of the regulator. Consequently, as with the other stakeholders, the relationship between a bank and its regulators is governed by many legal rules and regulations, but also by ethical concerns that form part of the contract. In particular, when implementing a policy for sustainable action, banks need to take their higher-risk carbon intensive business clients into account.

*Shareholders* in a bank deserve special attention among the stakeholders. This is because, according to the theory of corporate finance, the shareholders are the residual claimholders. Shareholders are paid only after everyone else is paid, which means that they bear the ultimate risk of the continuity of the enterprise. In a bank, shareholders bear the ultimate risk of any unforeseen losses that occur due to borrowers not being able to pay back their loan. When a policy for sustainable action in the banking industry aims to be acceptable to all parties in the nexus of contracts, it must also take the position of the shareholders into consideration. Before discussing this further, however, we first need a more precise understanding of the shareholder's interest. This we will do in the following steps.

First, we must clarify what shareholder value means. In short, the value of a company's shares is the present value of the expected free cash flows from now into the far future (Vermaelen 2009, 207).[3]

Second, we note that this understanding of shareholder value is based on the concept of free cash flows that a company is able to generate.[4] We further note that focusing on cash flows does not imply the maximisation of profits (Vermaelen 2009, 209). As can be seen from the definition of free cash flows, shareholders are residual claimholders: they are paid after everyone else is paid, whereas profit is a concept from accounting that potentially contains non-cash items and does not necessarily tell us anything about the value of the firm to the shareholders. In summary, the value of a company to the shareholders consists in the expectation about the free cash flows the company will be able to generate in the future, and determining the present value of these cash flows on the basis of an expected rate of return that reflects the risk that the company will fail to generate the cash flows.

Third, estimating the cash flows of a company in the future is a subjective matter. It requires in-depth knowledge of those aspects that drive the value of the business of the company. This means that estimating the value of a company requires a long-term view, which in practice typically means that a company's shareholder value can be explained by cash flows that occur in the range of 8 to 10 years (Koller *et al.* 2010). In determining the shareholder value as the present value of the expected free cash flows, a long-term view on the company must be taken, and future cash flows have to be estimated based on the company's plans and the development of the value drivers of the business over time.

In fact, the contract between a company and its shareholders involves mostly ethical features. The explicit legal part of the contract consists in certain voting rights. A small shareholder, however, by exercising these rights, can only marginally influence her position. This means that the contract between a company and its shareholders mainly consists in the implicit assumption that the company will provide the shareholders with a return that reflects the risk they take by investing their money. Note that I do not claim that the company should maximise its value to the shareholders, which is a customary phrase in the corporate finance literature (Vermaelen 2009). There are two reasons why it is rational for corporate managers to meet their shareholders' expectation of realising a reasonable return on their capital.

First, if corporate managers pursue strategies that benefit other stakeholders but not the shareholders (e.g. spending large amounts of free cash on philanthropy), then the stock price of the company will decrease, as shareholders will look for other opportunities to realise their expected returns. For the shareholder, who is pursuing a return on her investment, it seems irrational to invest in a company with a particular risk profile, when she could also invest in another company with the same risk profile against a significantly better return. Since her influence as an individual shareholder is very limited, she might sell her stock. A low-valued stock hampers a company in its growth opportunities.

Second, as we have assumed, companies operate in global, competitive markets. The concept of global competition has resulted in a market for corporate control also developing. That is, the undervalued stock of a company – the result of managers not living up to the requirement to deliver an appropriate risk-return relationship for their shareholders – attracts the attention of competitors, who may issue a bid for the stock. When these competitor-shareholders succeed, they replace the incumbent management with managers who are set to deliver shareholder value involving a lower cost base. Jensen and Meckling (1976, 308) talk about a reduction of "agency costs". Hence, it is in the interest of corporate managers who pursue a course of sustainable action that the policy they propose is beneficial to the shareholders.

So far, we have considered the environment in which bank managers operate, which involves at least five stakeholder groups. We are now ready to analyse the main difficulty in implementing a policy for sustainable action.

## 8.3 Challenges involving one's commitment to a sustainable policy

It follows from the contractual approach that a proposal for sustainable action in the banking industry that aims to be realistic enough to be accepted, must take care of the interests of all the stakeholders. In summary, these interests are: retail clients want their money back at all times against a competitive interest rate, and do not want to be associated with certain non-sustainable business activities; business clients only accept offers for a financial product at competitive rates; employees want to be relevant to their stakeholders and keep their job; the regulator's interest is to maintain a safe, reliable financial infrastructure by preventing bank defaults; and shareholders want a reasonable return on their investment as compensation for taking the role of residual claimholder. The interests of the stakeholders determine the interests of the bank and vice versa. That is, the stakeholders represent forces that limit the degrees of freedom bankers have for choosing courses of action that are in the interests of the bank. Simultaneously, actions taken by bank managers influence the ability of the stakeholders to pursue their interests. We will now discuss the further consequences for the interests of the stakeholders of a policy that promotes sustainable action.

As Humanity's Challenge is based on the Paris Agreement (UNFCCC 2015), we will discuss sustainable policies in the banking industry in the light of their effect on the reduction of greenhouse gases. The World Business Council for Sustainable Development (WBCSD) and the World Resources Institute (WRI) have developed a vocabulary (WBCSD and WRI 2004) that clarifies the control that companies have over greenhouse gas emissions. Direct greenhouse gas emissions occur from sources that are owned or controlled by the company, for example, emissions from combustion in owned or controlled boilers, furnaces, vehicles, or emissions from chemical production in owned or controlled process equipment (Scope 1). Scope 2 emissions are related to the generation of purchased electricity by the company. Scope 3 emissions are a consequence of the activities of the company, but occur from sources not owned or controlled by the company. Examples of Scope 3 activities are extraction and production of purchased materials, transportation of purchased fuels, and use of sold products and services.

At first sight, banks have two options for sustainable strategies. The first option is that they limit their own environmental footprint, through reducing paper usage, limiting air and car travel, using climate neutral buildings, buying electricity from renewable energy sources (Scope 1 and 2). The second option is that they somehow approach their retail and business clients with the aim of reducing *their* environmental footprint (Scope 3). As the greenhouse gas emissions of banks are mainly driven by their employees, and not by industrial processes, the first option ought to be implemented; however, its contribution to meeting Humanity's Challenge is of relatively minor significance. We thus focus on the second option as a bank's sustainable policy, which potentially contributes to a much larger extent to mitigating climate change.

This policy entails bankers shifting their priority towards increasing lending volumes with business clients who want to invest in sustainable business processes or have already done so. Correspondingly, bankers will want to decrease their lending to clients who have no plans whatsoever for a sustainable course of action. Given their constitution by stakeholder interests, this means that bankers must assess the impact of such a policy on their stakeholders. I will now look at the stakeholder groups in order to assess the conditions under which the proposed policy would be acceptable to them.

Before considering the points of view of the stakeholders, however, two remarks have to be made. First, traditionally, there has been a relationship between the *revenues* that banks realise on the loans they provide to business clients and the *risks* that banks are subjected to in this activity. The economics of banking prescribes that the revenues banks realise on their medium- to high-risk business clients must compensate for the costs caused by taking these risks (i.e. borrowers who do not pay back their loan). In addition, from these revenues retail customers must be paid a return in conformity with the market, and the costs of maintaining the banks' own organisation (i.e. employees) and infrastructure must be paid. This means in broad terms that when they lend to their business customers, the risk-return relationship for banks can be characterised by low risks that require modest returns, and high risks that should be compensated by high returns.

Second, as a result of changes in bank regulation in recent years (i.e. the implementation of the Basel II guidelines by the coming into force of the Capital Requirements Directive (CRD) II in 2006 throughout the European Union), the following developments have taken place concerning the risk-return relationship in the lending services of banks. When calculating the capital needed for a bank to maintain its lending practice, it has become a legal requirement to take into account the risk profile of its clients and products. As article 84.2 (b) of CRD II states:

> Internal ratings and default and loss estimates used in the calculation of capital requirements and associated systems and processes play an essential role in the risk management and decision-making process, and in the credit approval, internal capital allocation and corporate governance functions of the credit institution.
>
> (CRD II, 2006, 59)

Specifically, following the implementation of quantitative credit risk models, the variation ('granularity') in the risk profiles of clients has increased significantly compared with the previous regulatory regime. In 2014, the Basel III regulations came into force (CRD IV 2013). One of the main consequences of this regulation is that the connection between the risks a bank assumes in its lending operations and the return on their investment for the bank's shareholders has become even tighter. The main difference with the traditional concept of Return on Equity (ROE), defined as Net Profit as a percentage of shareholders' Equity, is that Equity has been replaced by $c \times RWA$ for internal capital allocation purposes, whereby the constant c is chosen to be equal to the bank's Core Tier 1 ratio, and RWA stands for Risk Weighted Assets – that is, the risk weighted exposure of the bank towards its lending customers. This leads to the following formula being used as the basis for defining ROE under Basel III: ROE = Result/Capital = Result/ ($c \times RWA$). Through the ratio between result and risk, the return on the shareholders' capital is now directly related to the revenues banks receive for extending their services and the risks they incur by doing so.

We can now assess the consequences of the bank's policy to increase its lending volumes with business clients working on reducing their greenhouse gas emissions. As the regulators indicate, a differentiation in risk profile is likely to emerge between carbon intensive companies that strive for a green future and companies that do not take the required greenhouse gas reduction measures in time. Assets of the latter group might need to be written off before their economic life ends, potentially causing an accelerated depreciation of the bank loans that finance the respective companies. For the continuity of the financial system, it is thus in the regulator's interest that banks anticipate the change in risk profile of their clients and start migrating to clients with sustainable policies sooner rather than later. However, the bank's decision to pursue a sustainable policy changes the equilibrium between the stakeholders, as expressed by the ratio between result and risk.

Let us look at what this change entails. The risk that banks run on business clients committed to mitigating climate change decreases. Since these clients anticipate a shortage of resources in the near future, they are likely to be among the companies that will survive the project of mitigating climate change, compared with companies that do not take any measures. Following the market's transparency, however, clients with a sustainable policy will demand lower interest rates in return for their lower risk profile, causing a corresponding reduction of the bank's revenues. The question, then, is: Will the reduction in risk compensate sufficiently for the reduction in revenues, leaving the equilibrium between result and risk in place? Note that fully answering this question requires further empirical research, which falls outside the scope of this book.

However, I argue that the decrease in revenue will potentially be larger than the reduction of the credit risk. In fact, the phenomenon of reduced revenues from effective sustainable companies has already been observed in the financial markets. Companies with a good performance on sustainability borrow against better conditions than companies that are not yet taking robust measures (Kim, Surroca, and Tribó 2014). On the other hand, a decrease of the risk factor is theoretically possible. Credit risk entails three components, of which one will partly decrease (that is, the company's probability of default). However, to my knowledge, the effect of a company's sustainability performance on its credit risk has not yet been investigated empirically. My hypothesis, however, is that the ratio of result to risk decreases as a result of the revenue and risk changes for business clients caused by the bank focusing on clients with a good sustainability performance. In that case, the other stakeholders will be affected.

To see why this hypothesis is plausible, recall that the bank operates under conditions of full competition. This implies that the reduction in revenues cannot become the burden of one of the stakeholder groups alone. As indicated, retail customers may look for another bank if they find out that they are going to pay the price for the bank's sustainable policy in the form of reduced rates on their deposits. In a similar way, employees whose salaries may need to be reduced following the need for an overall cost reduction exercise could start looking for another employer. Finally, the shareholders, who are confronted with a lower return if the other stakeholders do not move, may sell their shares, leading to a lower market value of the bank. Such lower market value could then become a signal to the market of corporate control of higher than usual agency costs, which in turn may threaten the bank's independence.

Deciding in favour of a policy of stimulating lending to companies with a good sustainability track record may thus send a signal to the financial markets that there is a price to be paid in terms of a reduced ratio between result and risk. Given the conditions under which banks operate, it follows that this burden cannot just be shifted to one of the stakeholders, as it will not be rational for them to accept a unilateral impairment of their interest. As bankers and stakeholders have entered into their relationship under conditions that both have accepted, it would also be unethical to impose the burden of the sustainable policy on one of the stakeholder groups. The decision to implement the sustainable policy may lead to the need to find a *new equilibrium* between the bank and its stakeholders as expressed by the result-to-risk ratio. The consequence of the reduced rates that business clients demand for their contribution to mitigating climate change could therefore be that retail clients have to accept lower rates on their deposits, that employees have to accept lower salaries, and that shareholders have to accept a lower return.

In the agent's experience the following could then happen. While she is in social cooperation with other agents about details of practically implementing the policy, the agent becomes aware that her efforts are being frustrated. From the behaviour of her colleagues, she may infer that the original optimism about the policy has turned into silent scepticism. Many of the policy's original supporters seem 'busy' with other priorities. For the agent, it feels as if the organisation has expressed a pocket veto over her policy. Momentum is lost, the agent's self-control is exhausted, and the policy remains on the shelf waiting for roll out. Again, future generations end up at the bottom of current people's priority lists.

Deciding in favour of a sustainable policy may thus require an action that prima facie does not follow from the instrumental coherence required for implementing the policy. Although we suppose that the agent – our bank manager – has developed the capabilities that are required to pursue the implementation of the policy, she may not be equipped to take the moral action of entering into discussions with the stakeholders with the aim of achieving a new equilibrium, as expressed by the lower result-to-risk ratio. Since the stakeholder relationships are both legal and ethical in character (see §8.2), simply upgrading the agent's negotiation skills will not work. The agent needs to reach a renewed understanding with the stakeholders about the ethical role of the bank in society, which entails biting the bullet of agreeing with the stakeholders about their share of the price to be paid for the sustainable policy. Should the problem of redefining the stakeholder relationships not be solved, stalemate is likely to occur. Consequently, the change to a sustainable lending practice will fail.

The bank manager faces the situation of being confronted with an existential choice. Either she remains silent towards the stakeholders with the likely effect that severe inaction remains. Or, she raises the alarm and uses her position to make the stakeholders aware of what is going on, with a view to making them move from their position in the discourse about the sustainable policy. This, however, may bring her into a similar situation as the purchase manager (see §3.5). The stakeholders, who basically define the role of the bank manager, could begin redefining her role (e.g. giving her another job with fewer possibilities to influence any sustainable policy), or they could consider opting out of the relationship altogether. Both possibilities entail significant risks for the agent, of which she perhaps was not aware when she committed to taking the action to promote the sustainable policy.

In summary, by considering the potential impact of implementing a sustainable policy on the stakeholders, it turns out that a new equilibrium among them might be required, as indicated by a decrease in the result-to-risk ratio. However, establishing this new set of stakeholder relationships may be too heavy an on-the-spot burden for the agent. That is, although we assumed the agent to have overcome the ex ante burdens (that is, possessing the agential capabilities, and being supported by appropriate institutional policies), her sustainable action may bring her into a situation that she may experience as being similar to waging or ending a war. As a consequence, the agent's motivation may become severely frustrated, leading to inaction and stalemate within the institutional set-up of stakeholder relationships. It is on the basis of these considerations that we have identified the capability of the second-order commitment that agents need to develop in order to overcome this motivational problem (see §7.4). In the next section, we will continue with the case study by showing how such a second-order commitment could be exercised and how it could potentially change the banking sector.

## 8.4 Sustainable action and potential change in the banking sector

In April 2017, ING Bank introduced a lending product that couples the interest rate on a loan to the business client's sustainability achievements (ING 2017; Philips 2017; Schoenmaker and Schramade 2019, 295). Companies can use these loans for their corporate purposes in general, not just for environmentally friendly projects. When the borrower's sustainability performance improves, the interest rate decreases. On the other hand, when the borrower's sustainability performance deteriorates, she will pay more interest. ING has labelled this loan a 'sustainability improvement loan', thereby emphasising that the business client's commitment to do better on sustainability is most important (ING 2019).

The product makes use of the assessment of a company's sustainability achievements by an independent, specialised rating agency. Based on a questionnaire regarding, for example, the company's greenhouse gas emissions, the company's social practices and the way in which it is managed, the agency produces a report with an overall sustainability score for the company.

Typically, the sustainability score is a natural number between 1 and 100. The better the company performs on sustainability, the higher the score. By implementing sustainable policies – for example, reducing its water consumption, or improving the health and safety of its employees – a company can improve its overall sustainability score. As mentioned above, the sustainability rating agency is independent, meaning that neither the company nor the bank can influence the final sustainability score; it is determined by the analyst working for the rating agency.

In the loan agreement between the borrower and the bank, reference is made to the improvement of the sustainability score by the company. First, the company acknowledges the sustainability score as representing the actual situation in the company with regard to the status of its sustainable policies, actions, and results. Second, the company undertakes to inform the rating agency during the life of the loan about its sustainable policies, actions taken, and external recognition received – to the extent that this is reasonable and practicable. Third, within the clause that formulates the pricing of the loan, it is stated that if the company improves its sustainability score by x points (say from 60 to 70), then the interest rate will decrease by y per cent (say from 1 per cent to 0.95 per cent). Conversely, if the sustainability score for whatever reason decreases by x points (from 60 to 50), the company will incur an increased interest rate of y per cent (from 1 per cent to 1.05 per cent). Finally, in order to prevent a situation in which the company benefits from the lower interest rate just by disclosing a few of its policies to the rating agency, the bank insists that the sustainability score when signing the loan agreement cannot be older than six months. In other words, the sustainability improvement loan is there to incentivise concrete sustainable action taken by the business client, and not just the disclosure of documents to the rating agency.

Since the introduction of the sustainability improvement loan in 2017, three developments have taken place. First, the product concept has been extended. Businesses and banks have discovered that, for some companies, sustainability achievements can be approximated by focusing on specific indicators, such as greenhouse gas emissions, renewable energy and water consumption, rather than the sustainability score by the rating agency. For these key indicators, quantitative goals beyond the term of the loan are being defined and progress is measured annually. Similar to the case of the sustainability rating agency, the company's progress regarding meeting the goals is verified by an independent auditor (DSM 2018, EDF 2018).

Second, the market has renamed the sustainability improvement loan as the 'sustainability linked loan'. In March 2019, the Loan Market Association (LMA)[5] adopted the so-called Sustainability Linked Loan Principles (SLLP). As the LMA states:

> The goal of the SLLP is to promote the development and preserve the integrity of the sustainability linked loan product by providing guidelines which capture the fundamental characteristics of these loans. In doing so,

> the purpose of the SLLP is also to promote sustainable development more generally. The SLLP are voluntary recommended guidelines, to be applied by market participants on a deal-by-deal basis depending on the underlying characteristics of the transaction.
>
> (LMA 2019)

The SLLP are meant to facilitate the borrower's achievement of ambitious, predetermined sustainability performance objectives. This means that companies are encouraged to make public their sustainability strategy against the background of their overall company strategy, and to report annually about the progress they are making. Sustainability objectives should be based on performance levels not older than 6 to 12 months. Finally, the SLLP recommend independent review of the company's sustainability achievements.

Third, market research firm Bloomberg has reported a strong growth of sustainability linked loans worldwide. From the beginning in 2017, the global volume of sustainability linked loans that were provided by a group of banks (i.e. 'syndicated') increased during 2018 by 677 per cent to $36.4 billion within the total market for sustainable debt, which grew by 26 per cent to $247 billion (Bloomberg 2019). To put these figures into perspective: the global syndicated loans volume reached $5.28 trillion in 2018 (Dealogic 2019), which means that sustainability linked loans make up 0.7 per cent of the entire syndicated loan market.

On the basis of the investigation of this book, we can make the following observations about the sustainability linked loans that were introduced as part of ING's sustainable policy. First, they make use of the characteristic of intentions as conduct-controlling pro-attitudes that exhibit inertia and function as inputs for further practical reasoning (see §5.2). Once the company has expressed in public its commitment to improve on sustainability, agents within the company tend to take some sustainable actions, or at least begin to develop plans to do so. This means that the company's sustainability score, all other things being equal, will increase, or will remain stable after signing of the loan. In other words, if the company is committed to taking sustainable action, under normal circumstances the probability will be low that the company will need to pay a higher interest rate due to a decrease of its sustainability score.

Second, expressing the company's commitment in public and relating it to an increase in the sustainability score is in conformity with all of the proposed informational institutional interventions (see §7.3). Expressing its commitment to improve on sustainability allows the company to inform the public (outside and inside) which sustainability challenges are relevant for them. Setting the goal in terms of x points improvement in the company's sustainability score provides the opportunity of breaking down the high-level goal of meeting Humanity's Challenge into sub-goals, which can be made actionable for the employees – that is, brought within the phenomenology of their agency (see §4.3). The inclusion of the x points improvement of the company's sustainability score in the loan agreement has two effects. It strengthens the company's determination to execute, and it prompts the organisation to deliver this improvement; it is a pledge to take sustainable action, which prevents agents from nonreflectively reconsidering their intentions. Finally, the annual independent review of the sustainability score by the rating agency means that the company obtains specific feedback with regard to the progress made, feedback that is essential for overcoming any intention-belief incompleteness (see §5.4).

Third, the sustainability linked loan is an example of an institutional intervention by means of psychologically informed nudging. In commercial enterprises, it is the role of employees working in the finance department to arrange the company's finances at the lowest cost possible. Company treasurers are the natural counterparts for bankers to establish a commercial relationship with. Since the interest rate on the loan decreases if the sustainability score goes up by x points, treasurers have a non-moral motive for promoting sustainable action within the company. The improvement of the sustainability score is in the direct interest of the treasurers, who do not necessarily need to be motivated to take sustainable action. In other words, these employees cannot reasonably deny that there is something in it for them if their employer lives up to her commitment to contribute to meeting Humanity's Challenge (see §1.1) and the banker proposes to advance a sustainability linked loan to the company.

Fourth, let us review to what extent the policy of the sustainability linked loan is aligned with the bank's stakeholder interests. For retail clients, the policy has no financial consequences; at the same time, retail clients can claim that their deposits allow a tangible reward to be given to businesses for taking sustainable action. Employees and regulators will usually appreciate the policy for the similar reason that in this way they make a concrete contribution to meeting Humanity's Challenge without paying a price for it. As it has been designed, the sustainability linked loan will be acceptable to the business client from both a moral and an economic point of view. That is, given the weak motivational force of the moral motive for meeting Humanity's Challenge (see §1.4), the economic motive of obtaining credit at the lowest possible cost is now aligned with the moral motive of promoting sustainable action.

Finally, we should consider how the sustainability linked loan affects the interests of the bank's shareholders. As already explained, this loan is based on the assumption that the slight decrease in the interest rate (i.e. Result) will be accompanied by a slight decrease in the probability of default for the portfolio of companies that achieve significantly better on sustainability (see §8.3). The decrease in probability of default, all other things equal, would then lead to a decrease in Risk Weighted Assets (RWA). The overall return for the shareholders on their equity – that is, ROE = Result/(c × RWA) – could then remain stable or increase. However, a slight decrease in ROE would also be possible. Regarding the impact of improved sustainability scores on a company's probability of default, more empirical evidence is needed to reach more definitive conclusions. Therefore, for the other stakeholders, the sustainability linked loan has positive implications, whereas this remains uncertain for the shareholders until empirical studies have been published.

Nevertheless, the above-mentioned developments following the introduction of the sustainability linked loan could be interpreted as a sign that other banks and corporations have adopted the concept quickly. The agent who promoted this sustainable bank policy needed to reach a renewed understanding with the stakeholders about the ethical role of the bank in society. This entailed agreeing with the stakeholders about their share of the price to be paid. Although it is still too early for final conclusions, it seems that the sustainability linked loan has led to such new equilibrium being found in the ethical relationships between banks and their stakeholders. The adoption of the SLLP as a standard and the growth of the volume of sustainability linked loans show that the bank's decision to proceed with its sustainable policy and introduce this product has elicited potential change in the banking sector – that is, change in the sense of integrating a tangible reward for taking sustainable action into the daily practice of servicing business clients.

Could this mean that we are making progress with living up to Humanity's Challenge? Perhaps a tiny bit. However, given the continuing year-on-year rising global greenhouse gas emissions, there is no reason whatsoever to be satisfied. The motivation problem, as it has been investigated in this book, remains unsolved, and formidable challenges lie in front of us. One of these is the lack of belief that one will follow through, or intention-belief incompleteness (see §5.3). Formulating and agreeing the goal of improving a company's sustainability score by x points could lead to the belief, among the agents whose task it is to make it happen, that, although one is committed, one simply does not believe that the organisation will be able to take the required actions. Another challenge is that companies still want to take instrumental sustainable action only (see §1.2). In other words, the incentive of y per cent on the interest rate is more important than reaching the goal of an improved sustainability score, let alone meeting Humanity's Challenge. In deciding about and working on the goal as it is specified in the sustainability linked loan, there still seems to be an element involved that can be captured by the phrase "What's in it for me?" (see §1.1).

Therefore, we need to find another, stronger motive that keeps us on track with contributing to meeting Humanity's Challenge. This is the motive that we will find when we dig into the concept of moral corruption, the topic of the final chapter of this book.

## Notes

1 I am aware that some readers may find too many details in this chapter; others may criticise that I have omitted crucial ones. To both I apologise beforehand.

2 I do not discuss other stakeholders such as NGOs, unions, governments, and society at large, since these institutions tend to influence the behaviour of bank

managers through one of the other stakeholders. They either influence the public in general (i.e. retail and business clients), the employees, the regulators, or the shareholders to take care of their interests.

3 In mathematical terms, when we will receive a payment F in n years from now, and i is the expected interest rate, then the worth P of this payment in today's money ("present value") will be: $P = F \times (P/F) = F \times 1/(1 + i)^n$.

4 Investopedia (2019) provides the following definition:

> Free cash flow represents the cash a company generates after cash outflows to support operations and maintain its capital assets. Unlike earnings or net income, free cash flow is a measure of profitability that excludes the non-cash expenses of the income statement and includes spending on equipment and assets as well as changes in working capital.

5 The Loan Market Association's aim is:

> to encourage liquidity in both primary and secondary [syndicated] loan markets by promoting efficiency and transparency, as well as by developing standards of documentation and codes of market practice, which are widely used and adopted. Membership of the LMA currently stands at 638 organisations and consists of banks, non-bank investors, borrowers, law firms, ratings agencies and service providers.
>
> (LMA 2016, 3)

# 9 Moral corruption and the pathway for humanity

## 9.1 Introduction

In Chapter 7, we discussed the pathway for individual agents who require certain agential capabilities in order to take sustainable action, and pathways for institutions that suggest interventions for developing these capabilities. In Chapter 8, a case study was elaborated to show how making a second-order commitment can be understood from the perspective of the agent. Now, as a final step, we turn to discuss the problem at the level of *humanity* that the lack of progress with mitigating climate change could be the result of our failure as individual and institutional agents to make the commitment to take up Humanity's Challenge. More specifically, we will consider in this chapter the problem of morally corrupt agency, which I tentatively understand as follows: We accept Humanity's Challenge as a normative requirement that applies to the collective of human beings; however, as *individual* and *institutional* agents we structurally persist in non-sustainable behaviour. In relation to this problem, the questions to be discussed are: What implications would it have for our practical understanding of ourselves as human beings if we were to continue persistently with our non-sustainable behaviour? And, how can we make an understanding of ourselves as human beings who strive for a eudaemonistically good life compatible with our tendency to fail to see Humanity's Challenge as a normative reason for us as *individuals* to take sustainable action?[1]

I argue for the thesis that human beings ought not to engage in morally corrupt behaviour lest they suffer from living with a divided practical identity for which they are morally responsible. This leads me to claim that, in addition to the weak motive (see §1.4), human beings have an indirect motive for making a commitment to taking sustainable action, while continuing to be a unified moral agent, and while also remaining a candidate for consideration as a person with moral integrity. To arrive at this thesis, my argument proceeds through the sections of the chapter as follows.

Up to this point, we have spoken about Humanity's Challenge as a challenge for 'humans' – that is, for the totality of human beings (see §1.3). However, 'humanity' can also be understood as the generic concept of what is human that consists of a description of what makes human beings human. Therefore, we start the chapter with clarifying my understanding of 'humanity'. We continue to elaborate Stephen Gardiner's notion of moral corruption in the context of taking action in order to mitigate climate change. Gardiner's formulation refers to Kant's *Groundwork for the Metaphysics of Morals* as follows:

> Moral corruption is (a) a tendency to rationalize, which (b) casts doubts on the validity and strictness of moral claims, by (c) seeking to pervert their status and substance, and in doing so (d) aims to make those claims suited to our wishes and inclinations, and (e) destroys the characteristics in virtue of which we respect them (e.g. what Kant calls their "dignity").
> (Gardiner 2011a, 307)

In Gardiner's view, this phenomenon should be seen as the root cause of an agent's attitudes of complacency and procrastination when it comes to taking sustainable action, even when such action is supported by the agent's moral judgement (Gardiner 2011a, 45). We will discuss Gardiner's formulation of moral corruption and its reception in the literature (§9.2).

On the basis of the conceptual-phenomenological analysis of morally non-trivial actions, as performed in this book, I choose the following formulation of moral corruption, which is more specific than Gardiner's. As in the tentative formulation I offer above, morally corrupt agents accept Humanity's Challenge as a normative requirement that applies to human beings collectively.

**_Moral Corruption_**

*Is (1) pursuing a strategy, (2) with the objective of keeping up one's current, convenient status quo, (3) of deliberately and persistently failing to commit to taking sustainable action or of failing to take measures that will foreseeably be necessary for one to be sufficiently motivated to keep one's commitment to taking sustainable action.*

Thus, I conceive of moral corruption as something that agents actually do. This implies that agents are free to act differently; their agency is not impaired. And since our actions constitute, at least in part, our practical identities, moral corruption has an effect on us. I understand moral corruption in relation to Kant's "radical evil". After elaborating my definition of the concept, I compare my account with the accounts of other authors. We then continue by analysing how we can understand moral corruption as a kind of self-deception (§9.3).

We ask: What is problematic with morally corrupt agency? Apart from the fact that Humanity's Challenge will not be met if human beings display en masse this form of agency, what is the problem with the effect of moral corruption on *us*? If we understand ourselves as beings who are constituted by what we do, what would be the effect on our practical identity of following a morally corrupt strategy? I will argue that this kind of agency leads to a fragmentation of our practical identity that curtails our agential capabilities as we exercise them every day in social institutions. Morally corrupt agents may not fail on deliberation about options for action, including sustainable actions; importantly, however, they fail to make any commitment to taking sustainable action. Consequently, they do not enter the executive phase. They believe that their course of action renders their lives comfortable. I argue that this belief is tragically false, since the schizophrenia of her practical identity, caused by her morally corrupt conduct, effectively prevents the agent from achieving the end ("of keeping up one's current, convenient status quo") she has set for herself (§9.4).

We then move on by asking whether we can understand ourselves as persons of integrity if we engage in morally corrupt agency. On the basis of a concept of integrity understood as a virtue, I conclude that our integrity is at risk when we consistently fail to commit to taking sustainable action whilst attempting to live an active life as a present moral agent. Persistently not taking up Humanity's Challenge casts doubts upon our integrity, even though we still believe we can be regarded as possessing this virtue. Moreover, we are ourselves responsible for this lack of integrity (§9.5).

Finally, I will connect this conclusion with the argument that our self-understanding as self-conscious agents, and as participants in social-communicative relations with other human beings, is fundamental for us as human beings. I argue for the above-mentioned thesis that we ought to resist moral corruption by intending to take up Humanity's Challenge. By considering moral corruption in relation to our practical self-understanding, I argue that we find a new normative reason to accept that taking sustainable action in order to mitigate climate change is good for future generations *and* good for us (§9.6).

## 9.2 'Humanity' and Gardiner's concept of moral corruption

Our tentative understanding of moral corruption is that we accept Humanity's Challenge as a normative requirement that applies to the *collective* of human beings, but that, as *individual* and *institutional* agents, we structurally persist in non-sustainable behaviour. Since we want to investigate the implications of this kind of behaviour for our practical understanding of ourselves as human beings, we begin this section by clarifying how we understand 'humanity' as referred to in Humanity's Challenge. We then elaborate the concept of moral corruption, as Gardiner has formulated it. Gardiner's formulation is based on Kant's *Groundwork for the Metaphysics of Morals*, where Kant discusses, in the last paragraphs of the first section, how the human being feels in herself a counterweight against the commands of duty, which reason represents to her (Kant 2002, 20). In recent years, Gardiner has developed an understanding of this concept in the context of sustainability. Subsequently, we discuss the reception of Gardiner's concept by other authors.

In the remainder of this chapter, we will develop this concept further, on the basis of the work that has been done in this book.

Note that we have formulated Humanity's Challenge as follows (see §1.3):

> A basic question that faces us as humans, then, is whether we can meet the challenge of changing our way of life such that we live up to our commitment to do justice to the people who will live long after us by taking sufficient sustainable action to meet the target of "holding the increase in the global average temperature to well below 2 °C above pre-industrial levels ..., recognizing that this would significantly reduce the risks and impacts of climate change".

Clearly, humanity is understood here as the totality of human beings. That is, the challenge is relevant to each and every human being currently alive. Furthermore, I have argued that human beings currently living in the affluent countries form the moral community that ought to take up Humanity's Challenge and begin changing their way of life through taking sustainable action (see §1.5).

In his second formula of the Categorical Imperative in the *Groundwork*, Kant gives a fundamentally different interpretation of humanity when he states: "Act so that you use humanity as much in your own person as in the person of every other, always at the same time as end and never merely as means" (Kant 2002, 46–47). Kant refers here to humanity as the generic concept of what it means to be human that consists of a description of what makes human beings human. Allen W. Wood understands such a generic description of what it means to be human along the lines of Kant's own description of the three fundamental human predispositions, namely

> (1) animality (through which we have instincts for survival, procreation, and sociability); (2) humanity, through which we have the rational capacities to set ends, use means to them, and organize them into a whole (happiness); and (3) personality, through which we have the capacity to give ourselves moral laws and are accountable for following them.
>
> (Kant 2002, 47)[2]

Given these two formulations of humanity, it follows that Humanity's Challenge remains a challenge for the totality of human beings – that is, according to the first understanding. However, the question of how we practically understand ourselves as human beings if we persistently continue our non-sustainable behaviour must be seen against the background of the generic concept of what makes human beings human. This means that our investigation of moral corruption is about looking for possibilities to integrate into the human predispositions our persistently non-sustainable behaviour, considering that we have accepted Humanity's Challenge as a normative requirement that counts for the *totality* of human beings.

By analysing the way in which we generally do not live up to Humanity's Challenge, I argue for an understanding of the normative generic concept of the human being that follows from how we practically understand ourselves when we display morally corrupt behaviour. The thesis I argue for (see §9.1) states that we ought to stay away from moral corruption by making the commitment to take up Humanity's Challenge lest we suffer from living with a divided practical identity, for which we are morally responsible. Therefore, the investigation into moral corruption and our practical self-understanding leads to interpreting the human predispositions in such a way that the feature of taking sustainable action in order to contribute to efforts to meet Humanity's Challenge is good for future generations *and* good for current human beings.

Stephen Gardiner introduces the metaphor of a perfect moral storm to illustrate the way in which the problem of climate change has become a unique problem for humankind (Gardiner 2006). We have already seen how additional complexity is created through the mutual reinforcement of *global* problems, *intergenerational* problems, and *theoretical* problems (see §1.3). Gardiner argues that due to this cumulative moral complexity, we are vulnerable to the phenomenon of moral corruption. As the current generation of agents in affluent countries, hardly confronted with the consequences of climate change, we could easily "engage in *manipulative or self-deceptive* behaviour by applying one's attention selectively" (Gardiner 2006, 408, emphasis Gardiner). According to Gardiner, given the standard of living in the affluent countries (see §1.5), it may turn out to be convenient for us to simulate concern with the problem of climate change, whereas in fact we are simply exploiting our temporal position of living before the great climate catastrophe (ibid.).

Gardiner further develops the concept of moral corruption, while referring to Kant's *Groundwork* (see above), when he formulates that "moral corruption is (a) a tendency to rationalize, which (b) casts doubts on the validity and strictness of moral claims, by (c) seeking to pervert their status and substance, and in doing so (d) aims to make those claims suited to our wishes and inclinations, and (e) destroys the characteristics in virtue of which we respect them (e.g. what Kant calls their 'dignity')" (Gardiner 2011a, 307). In other words, the root cause of moral corruption lies in the fact that moral claims are in some sense uncompromising, and in particular do not seek to accommodate other sources of motivation, for instance the inclinations of the agent on whom the claims are made. Given this, agents have a tendency to develop rationalisations that chip away at the strictness and validity of their moral duties in order to reconcile them with their inclinations. In a worst-case scenario, these rationalisations incentivise some ill-conceived and potentially dangerous actions and provide agents with justifications or apologies for immoral behaviour (McQueen 2016, 4).

Gardiner continues to argue that it will be difficult to find an agent in social life who can be accused of moral corruption on the basis of this definition, since there are so many subtleties involved in the thinking and talking of human agents. Even agents who offer bad arguments or mistaken values for not undertaking sustainable action may do so in good faith, since they are unaware that they hold false beliefs. As we have seen in Chapter 4, agents may be misled by certain cognitive failures for which they cannot be held accountable. Gardiner thus concedes that "modes of moral corruption are so rich and varied, and because it is hard to isolate the morally corrupt from the merely mistaken, charges of moral corruption are difficult to prosecute" (Gardiner 2011a, 308). However, instead of identifying who is to blame for it, the more important question is, according to Gardiner: How can we fight moral corruption? Below, I will argue that, on the basis of the conceptual-phenomenological analysis of this book (see Part II), a clearer interpretation of moral corruption can be developed that could help in our attempts to overcome moral corruption.

Gardiner ends the defining part of his account by arguing how complex our task is to resist moral corruption. Relying on theories about global, intergenerational, and environmental ethics only, showing that these accounts are correct, and then assuming that humanity will convert to sustainable action probably would be naïve. Even if these theories were to be supported by human agents globally, the problem with motivation for sustainable action would still remain (see §1.3). Since human beings have difficulty with acting morally, to begin to meet Humanity's Challenge we cannot rely on agents in the affluent countries simply invoking the theories and applying them right away. The problem of our motivation for sustainable action being liable to moral corruption, therefore, needs a different treatment. Gardiner seeks a solution for this problem by drawing a parallel with the story of John and Fanny Dashwood in Jane Austen's *Sense and Sensibility*, which I will not elaborate here.[3] However, Gardiner concludes pessimistically that those who have the moral duty to mitigate climate change not only suffer little or no negative consequences from their failure to act, but they also stand to benefit from that failure, whereas the consequences of not acting are felt entirely by future generations (Gardiner 2011a, 336).

To date, Gardiner's position on moral corruption has been discussed in the literature only a few times. Peter Singer comments that Gardiner's perfect storm may be overly pessimistic. Back in 2006, Singer saw reasons for hope in the signing of the Kyoto Protocol by all the developed countries except the US and Australia, and the continuing discussions about further steps that need to be taken. Singer criticises Gardiner's account for being general and not explaining differences among the developed countries with respect to their contribution to the Kyoto process. Singer notes that due to the number of victims of Hurricanes Katrina and Rita in the US (2005), the so-called "climate loonies" in that country may have found a self-interested motive, rather than an ethical motive, to start to mitigate climate change. For that reason, they might escape Gardiner's perfect storm and its consequence of moral corruption (Singer 2006, 421–422).

Instead of moral corruption, Wouter Peeters *et al.* focus on what they call moral disengagement as the psychological mechanism through which people reconstruct Humanity's Challenge in order to evade individual responsibility for living up to it (Peeters *et al.* 2015, 3). They describe the mechanism of moral disengagement as: "to convince oneself and others that one's reprehensible conduct still falls within moral standards through changing the perception on one's actions and reconstructing the situation so as to reduce its moral intensity" (Peeters *et al.* 2015, 38). That is to say, by relying on one's moral disengagement one can resolve the inconsistency between one's moral standards and self-interested conduct by articulating reasons why one's reprehensible behaviour can be justified or excused as an exception to general normative rules. Peeters *et al.* provide various examples of how moral disengagement works: advantageous comparison, euphemistic labelling, minimising consequences, discrediting evidence of harm, and claiming ignorance (Peeters *et al.* 2015, 39).

The main part of the work of Peeters *et al.* (2015) is devoted to a discussion of Scheffler's phenomenology of agency (see §1.4) and how these examples of moral disengagement play a role in downplaying the moral responsibility of individual agents for taking up Humanity's Challenge. Their application of moral disengagement to Scheffler's phenomenology forms the basis for an understanding of the problem with motivation for sustainable action (see §1.5). Peeters *et al.* claim (2015, 100) that the prevailing liberal-capitalist worldview[4] encourages agents to believe that self-interested pursuits can be equated with wealth accumulation and consumption. In their view, climate change challenges the inviolable status conferred on materialistic freedoms, which requires agents to resort to moral disengagement in order to maintain their consumptive lifestyle without having to assume moral responsibility for harms that result from their behaviour (Peeters *et al.* 2015, 94). According to Peeters *et al.*, agents' propensity for moral disengagement can be tackled mainly by making people aware that it is a kind of (self-) deception. In that way, making agents aware of the mechanism of moral disengagement might reduce its persuasive force (Peeters *et al.* 2015, 117).

Below, I will propose that moral corruption should be understood as deliberately and persistently failing to commit to taking sustainable action (§9.3). I now argue that, for two reasons, the conceptual-phenomenological analysis of this book (i.e. Part II) could help to improve Gardiner's account of moral corruption. First, it seems difficult to apply the concept in an actual discourse between agents. How should we interpret the unavoidable "tendency to rationalise" in Gardiner's definition, and argue with other agents about whether they display such a tendency? In a discussion between agents in an action situation (such as the case of S who proposes that she and her fellow homeowners install solar panels, see §1.1), it seems unclear how the agents can determine whether they manifest this tendency so as to reach agreement on the question: Do you, do I, or do both of us display a tendency to rationalise of Gardiner's criticisable kind? And if the answer is affirmative, what can we do about it?

Although intuitively, we perhaps grasp what is meant by "the characteristics in virtue of which we respect them [i.e. moral claims]", in my view, just as with the tendency to rationalise, it will be difficult to bring this argument into an actual discourse between agents involved in some morally relevant situation. There does not seem to be a universally accepted description of the characteristics of moral claims that cause us to respect them. And, what does it mean that we respect moral claims? Would this imply that we act according to the norms we respect? As I have argued, the topic of this book is the observation that agents are not necessarily motivated to act according to a norm they have accepted (i.e. Humanity's Challenge – see §1.3). Even though Humanity's Challenge is respected as a moral claim that applies to the totality of human beings, this does not guarantee that as an individual we live up to this claim.

This brings me to the second aspect of Gardiner's concept of moral corruption that needs improvement. As Marcello Di Paola observes, the way it is defined, moral corruption is "corruption *of the understanding*, of the way we think and talk of climate change" (Di Paola 2013, 62, emphasis Di Paola). A consequence of Gardiner's definition is that it cannot be related to what agents actually do – that is, an action carried out that qualifies as a corrupt action. To be labelled a corrupt person presupposes an action that qualifies as corrupt, for example bribery or nepotism, for which the agent can be held morally responsible (Miller 2011, §2). In other words, the concept of moral corruption should specify those aspects of an action that cause it to be morally corrupt. From the concept of corruption, it follows that there must be someone who acts as the corruptor, and someone or something who is the corrupted – that is, someone or something undergoing the corrupting effect of the corruptor's action. It is important to note here, though, that corruptor and corrupted can be one and the same person, meaning that agents can corrupt themselves by acting in a certain way.

Gardiner's concept of moral corruption seems hard to use in the agent's reflections on her own way of acting. It seems difficult to be clear about what agents do wrong. The current definition does not allow agents who, in a reflective mode, want to trace where and when they ought to have acted differently in order to meet Humanity's Challenge, to pinpoint in which of the Rubicon action phases they should have made a different choice. Given the difficulty of connecting Gardiner's concept of moral corruption with an account of what it means to be a morally corrupt agent, I will now propose a reformulation of the concept, which is based on the phenomenology of motivation for sustainable action that has been developed in Part II.

## 9.3 Moral corruption based on the descriptive diagnosis

In this section, I propose an alternative formulation of the concept of moral corruption, one that is based on the conceptual-phenomenological analysis of Part II. Instead of deriving an understanding of moral corruption from accounts about corruption in general,[5] I take Kant's account of "radical evil" as a background for the concept that I am going to develop. I then formulate the new version of how moral corruption can be understood if we take into account the results from the descriptive diagnosis of the problems with motivation for sustainable action. As there are a few accounts by other authors that come close to our (new) understanding of moral corruption, we will analyse the relevant papers for similarities and differences to our definition. We continue by asking whether moral corruption can be considered as a strategy of self-deception. This analysis prepares us for the investigation of moral corruption in relation to our self-constitution as agents in §9.4 and our moral integrity in §9.5.

My provisional understanding of moral corruption is as follows: We do not make Humanity's Challenge a normative requirement that applies to us as *individual* agents, *and* we structurally persist in doing nothing about our non-sustainable behaviour (see §9.1). In other words, moral corruption is about *not* taking individual responsibility for doing something about this challenge, which has the structure of a tragedy of the commons (see §4.2). Whereas the agent accepts Humanity's Challenge as a normative requirement that applies to the *collective* of human beings, the new element is that the agent commits herself *not* to engage in sustainable action. Thus, moral corruption is something that agents actually do. As agents, in my view, make the commitment voluntarily to follow a morally corrupt strategy, this leads me to develop this sort of behaviour as an instance of Immanuel Kant's concept of "radical evil".[6] A summary of this concept now follows.

Kant argues that human beings have a propensity for evil. Evil, in Kant's view, is not just a superficial phenomenon; it is deep, as it rests on a radical decision made by the agent (Kleingeld 2007, 228). This decision is to put self-love above what morality asks from her. According to Kant, the deciding feature of "radical evil" is the agent's action to reverse the moral order of her motives by putting the principle of self-love above the universal law (Kant 1974, 21). Indeed, it is important to emphasise that Kant understands the agent's decision to follow evil as an *action*, just as it is understood in this book: voluntary and purposive (see §1.2). Since it is the agent's *free* decision to let self-love overpower the moral law, it is impossible to morally understand why the agent has decided this way. However, since the agent is free in her choice, she is solely responsible for her action (Kleingeld 2007, 229).

The use of "radical" must be understood in two ways. First, etymologically, "radical" is to be understood as root: the propensity for evil is rooted in human nature. There is no way to escape from this inclination; it cannot be eliminated from our existence. However, we can resist the propensity for evil; we are free to follow it or to resist it. Second, the agent's decision is also radical in the sense that it is far-reaching, for it applies to all maxims. Evil is not something external to the human being. All humans experience the temptation to subordinate what morality asks from them to the principle of self-love (Kleingeld 2007, 232). In subordinating the moral law and making her motives and inclinations of self-love a prerequisite for following it, the agent voluntarily and consciously chooses to take her inclinations to be reasons for action (Kleingeld 2007, 231–233).

Kant distinguishes three levels of evil. The first level is the one of inadequacy ("Gebrechlichkeit"), or, more precisely, weakness of will. One knows what one ought to do, one intends to do it, but one simply does not do it. One's inclinations are stronger than one's intentions. The agent chooses to let her inclinations determine her actions (Kant 1974, 22; Kleingeld 2007, 233).

The second level is impurity ("Unlauterkeit"). Here the agent adheres to a mingling of moral and immoral motives. One knows what the good action is; however, one does not do the action, since the moral law is not strong enough. The agent intends to do what the moral law prescribes – but only if other motives are supportive. In the agent's eyes, the moral law does not suffice for her to follow it (Kant 1974, 22; Kleingeld 2007, 234).

In the third case, called malice ("Bösartigkeit"), the agent chooses simply to put self-love above the moral law. There is no weakness of will or a mingled motivation, but just a pure selfish will. In this case, the agent has a bad will. Kant also names this form of evil corruption ("Verderbtheit") or perversity ("Verkehrtheit") (Kant 1974, 23; Kleingeld 2007, 234).

The account of moral corruption that follows starts from the (second) case of impurity, as this is the case that comes closest to what morally corrupt agents actually do. As I will construe them, morally corrupt agents have set themselves to pursue a strategy of keeping up their convenient status quo. They have freely chosen maxims of their will that are not universalisable, because if all human beings were to act on that maxim, climate change would grow out of hand and Humanity's Challenge would not be met. The point is that the agents I have in mind have decided on a principle to guide their actions that potentially could be at odds with the principle of morality that is, for Kant, the categorical imperative.

On the basis of the conceptual-phenomenological analysis of the Rubicon action phases, we have formulated the capabilities that agents need to develop if they want to meet the motivational challenges that come with taking sustainable action (see §7.2). Agents bear an individual responsibility for developing these capabilities, and I have made proposals for institutional measures that can enhance these capabilities in agents (see §7.2 and §7.3). Given the continuing problems with taking up Humanity's Challenge (see §1.5) and Gardiner's pessimism about our tendency to pervert morality (see §9.2), I will start with elaborating an understanding of moral corruption that benefits from the formulation of the agential capabilities, as developed in Chapter 7.

An understanding of the capabilities allows us to see the numerous ways in which agents can be ineffective in the sustainable action they intend to take. Knowledge of the capabilities of practical reasoning and judgement shows how agents can deviate from the course of sustainable behaviour when they stick to common-sense moral considerations or are not aware of their implicit assumptions about the conception of the human being they entertain.

Examination of the capabilities of social cooperation and recognition reveals the many situations in which sustainable action can be frustrated due to people failing to have a dialogue with each other. The same counts for the capabilities of the will, where we see agents fail by nonreflectively reconsidering their intentions or by being weak-willed. Otherwise put, the more we know about the motivational challenges, the more we become aware of the vast number of ways open to agents in affluent countries to excuse themselves for not taking sustainable action.

And to be fair, as the case study of the banker reveals (see Chapter 8), agents can indeed be confronted with complex situations, in which they are not sure about any form of backing by significant stakeholders. The investigation of the motivational problems with sustainable action emphasises the variety of situations where agents can take sustainable action, but where they need to take risks that may become unbearable for them. One example is the case of the purchase manager (see §3.5), where the CEO in fact saved her job after she was allowed to present her project to him, and he started to support it.

Nevertheless, with this nuanced understanding of the problems with motivation for sustainable action I do not want to suggest that nothing can be said about the problem of moral corruption that Stephen Gardiner has brought to our attention. On the contrary, the details of the motivational challenges help in locating the "radical evil" in our will that forms the major roadblock for taking up Humanity's Challenge. In my view, we cannot be at peace with ourselves if we do not fight and overcome this evil that is located in our will, which makes us feel comfortable with failing to meet Humanity's Challenge. As a potential way of coping with our distorted will, we must work on our practical self-understanding in order to see what is really required of *us* in order to meet our obligations to future people. That is, through developing an improved understanding of ourselves as agents, and of the effect on us of our morally corrupted will, we might become aware of what we can do to overcome the problem of failing to meet Humanity's Challenge. I will now argue for an understanding of moral corruption, as related to our ineffectiveness with mitigating climate change, which I see in the spirit of Kant's "radical evil".

The understanding of moral corruption I am looking for can be related to the powerful position of agents who are strongly willed in the sense of being capable of effective agency in complex action situations. However, these agents do not make a commitment to taking sustainable action, as they can exercise their power over the current poor, over nature, and over people who will be born in the distant future. If they wish to do so, these agents can persist in a course of passiveness when it comes to sustainable action, while simultaneously subscribing to the norm that agents in the affluent countries collectively ought to take sustainable action and believing that sustainable action can reverse the trends of climate change. However, nobody stops them from remaining passive, and their position as members of an affluent society provides them with an argument to do so.

Recall that the agent's commitment, as her existential choice to act sustainably, is crucial in making progress with taking up Humanity's Challenge. As I see it, an agent's commitment does not seem reducible to one of the required capabilities for being effective with one's sustainable action. Institutions can be ambitious with implementing policies; however, if large groups of individual agents simply do not cooperate, then such sustainability policies will fail. Alongside one's commitment (first-order) to take sustainable action, there is a second-order commitment required for solving the hard problems of, for example, acting in multiple stakeholder environments and developing the capabilities that are needed for effective sustainable agency (see §7.4). Therefore, I now propose the following understanding of moral corruption:

> ***Moral Corruption***
>
> *Is (1) pursuing a strategy, (2) with the objective of keeping up one's current, convenient status quo, (3) of deliberately and persistently failing to commit to taking sustainable action or of failing to take measures that will foreseeably be necessary for one to be sufficiently motivated to keep one's commitment to taking sustainable action.*
>
> *One knows that taking the sustainable action is morally required; however, one does not do the action, since the motivation of meeting Humanity's Challenge is not strong enough. This means that one does not make the plans that are necessary to overcome the motivational challenges of taking up Humanity's Challenge, in both one's individual and one's institutional roles.*

Overall, one can say that morally corrupt agents do their best to evade questions about their commitment to take up Humanity's Challenge. This means that morally corrupt agents do have moral prescriptions about taking sustainable action as part of their shared background assumptions. They know that they possess these assumptions somewhere in what we call their conscience. These agents, however, have decided to leave this part of their life world aside and not let it exercise any influence over the practical considerations that normally lead to making commitments. In short, morally corrupt agents are deaf to principal moral questions regarding sustainability and climate change. Since we are considering here agents who are capable of effective agency in complex action situations, these agents sense that their deafness to these questions may result in the irreversible corruption of their moral character and thereby the potential loss of a few moral virtues. This implies that moral corruption is not only a moral concept, but also a causal or quasi-causal concept: "an action is corrupt by virtue of having a *corrupting effect* on a person's moral character or on an institutional process or purpose" (Miller 2011, §2.1, emphasis Miller). Some of the terms used in this definition need further clarification.

First, moral corruption should be understood as pursuing a strategy, very much in the same way as one chooses a strategy in the context of decision theory. That is, one does not give up easily when attempting to achieve a certain goal that has a high utility. More specifically, if after having carried out action X, action Y is required to reach that particular goal, the agent will be strongly inclined to take Y, even when Y involves a morally blameworthy situation for the agent. It also means that one's endeavours are strategic in a Habermasian sense – that is, the agent has set herself unilaterally on a course of action to achieve a specific goal in such a way that she systematically thwarts any opportunities for others to give or withhold their consent – for example those people affected by the strategy that has been set (Bohman and Rehg 2014, §3.1).

Second, the strategy of morally corrupt agents is aimed at preserving their current convenient status quo in the high-income OECD countries. As Gardiner points out, it is an almost universal characteristic of decisions about ordinary consumption patterns that these are focused on the short and medium term at the most, and that their spatial scope is limited to local circumstances (Gardiner 2011a, 58–59). It means that people currently living in Western industrialised countries define their good lives in terms of certain social-economic, material conditions to which they want to stick. They seem to be of the opinion that, whatever happens, their wealth should be impacted only minimally, if at all. This is very much the way it was expressed by US delegates to the Earth Summit in Rio de Janeiro in 1992, when they insisted that "the American way of life is not up for negotiation" (Elmer-DeWitt 1992).

Third, the principal characteristic of morally corrupt agents is that they do not make a commitment to meet Humanity's Challenge. They do not form the intentions to take sustainable action. Morally corrupt agents know that intentions involve characteristic kinds of motivational commitments. As we have seen (see §5.2), intentions are conduct-controlling, in the sense that "if you intend to F at t, and nothing changes before t, then (other things equal) you will F" (Wilson and Shpall 2012, §4). The reason why they do not form intentions to overcome the motivational challenges of acting in the interest of future generations is that if one intends to do F, one is already in progress towards doing it. In that sense, deliberation is over: one has begun (Setiya 2014, §1). This is exactly what morally corrupt agents try to prevent.

Moral corruption should be understood in a similar way to Christine Korsgaard's elaboration of the four types of vices in Plato's *Republic* (Korsgaard 2009, 165). It does not make sense to go out on the street and look for primary examples of morally corrupt agents or to look in one's organisation for agents that more or less fulfil the characteristics of the definition. Rather, we should acknowledge that our motives for taking action to live up to our obligations to people living in the distant future might be morally corrupted in a way similar to how Plato describes people striving for a good – that is, aristocratic – constitution: they have to fight against the timocratic, oligarchic, democratic, and tyrannical tendencies that try to overrule the aristocratic governance of their soul. Many people living in high-income OECD countries have to face moral corruption as a serious obstacle to taking action to care for the distant future. Moral corruption is a 'shadow' for moral agents, with which they are confronted when they feel hampered in acting in the sense of taking up Humanity's Challenge.

The morally corrupt agent, in the sense of our definition, can be seen as structurally free riding on the efforts of others. Free rider behaviour means shifting the moral problem of not achieving the end of one's action to other people or future generations. One's failure to execute the act has negligible negative consequences for oneself, but potentially substantial negative consequences for current or future generations, should many other agents behave in the same way. Each of us exchanges a bit of effort or costs, for example by putting a device in our cars that reduces $CO_2$ emissions, in return for benefiting from some collective provision – in this example, clean air. We are free riders if we do not contribute through installing the device, yet still benefit from the clean air produced by others (Hardin 2003, §1). Perhaps this is the most severe temptation for agents not to take sustainable action: the low price they have to pay in terms of diminished self-interest if they do not act to prevent the potentially large impact that collective inaction will have for meeting Humanity's Challenge.

In sum, the morally corrupt agent has agreed that meeting Humanity's Challenge is the norm for humans collectively that justifies a sustainable course of action. She is aware that people in the affluent countries must take the lead in preventing global warming; she has even agreed to a proposal for sustainable action that has passed the deliberative phase (all-things-considered judgement). She just does not make the step to a commitment in the sense of binding her will to follow practical reason and to start executing the proposed action. She also does not commit to taking measures that are foreseeably necessary for her to be sufficiently motivated to keep her commitment to taking sustainable action. However, if the agent is committed to acting morally in many other situations, then how can she make her non-commitment to sustainable action compatible with how she understands her agency? How can she be morally corrupt regarding sustainable action on the one hand, and strive for a virtuous life on the other hand? What follows for her practical self-understanding as agent when she sticks to a morally corrupt strategy? I attempt to answer these questions in what follows in this chapter.

So far we have provided a positive description of moral corruption. But it would seem useful to give a negative description as well, in the sense of marking the concept off from related concepts that have been mentioned or proposed in the literature.

First, there is the question of whether moral corruption could be a form of akrasia. This means that the agent's action at t seems synchronous with her unfavourable judgement at t about that same action (see §6.3). Agents make their commitment, however, prior to acting – that is, at t-1. The morally corrupt agent has never formed an intention to take a sustainable action. Therefore, the agent does not even arrive at t, where a discrepancy between her action and her judgement could be observed. In that sense, morally corrupt agents do not fail to follow up on their intention to care for the distant future.[7] For the same reason, moral corruption cannot be reduced to weakness of will or procrastination, as these agents have not formed an intention on which they fail to act or in relation to which they display culpably unwarranted delay of their action.

Second, the root cause of our collective failure to live up to Humanity's Challenge could be named "the problem of failed intentions" (Brister 2013). Evelyn Brister argues that agents could become cynical as a result of their being caught between the seemingly imperceptible effects of their individual sustainable actions, and their judgement that governments are unwilling to implement the appropriate measures. As a way out of cynicism, Brister proposes to form informal non-governmental networks of individual agents, who support each other in reducing their contributions to climate change. This could take the form of, for example, lobbying for local policy changes and creating and adopting well-designed technologies (Brister 2013, 260).

I think that Brister's account differs from the concept of moral corruption as we have formulated it. In her view, agents have formed the intention to mitigate climate change; however, they face considerable difficulties in fulfilling them. As Brister's account takes the agent's intention to take sustainable action as a starting point, it cannot replace my account of moral corruption, which is grounded in the agent's non-commitment. In effect, from my perspective, Brister is proposing a remedy for one specific obstacle that agents may face as they try to implement their intention to live up to Humanity's Challenge. Generalising, we can see that her proposal requires what I call second-order commitment: commitment not to let one's first-order commitment be undone by the circumstances.

Third, one could ask whether moral corruption can be explained by referring to the agent's complacency. Michael Doan argues for complacency as a motivational vice in need of remediation, along with corruption and cruelty. In this regard, Doan's motivational vices are present when the ways someone has been constituted as a moral agent "prevent her from inquiring into, understanding, and responding well to a range of complex ecological and social problems" (Doan 2013, 3). In Doan's view:

> Agents … are complacent on climate change insofar as they are caught up in patterns of behaviour that express *settled expectations of self-sufficiency*. That is, I "lapse into" complacency by coming to expect that there is something *I* can do to become a good environmental citizen by improving my own behaviour and lifestyle, and when this expectation is so easily and frequently met by the people I interact with, and given the resources at my disposal, that it recedes into the background of my conscious awareness and becomes part of the way I expect the world to be.
>
> (Doan 2013, 11, emphasis Doan)

As moral corruption aims to maintain one's current, convenient status quo, it could be considered as common-sense complacency, understood as an uncritical satisfaction with oneself or one's achievements. Thus, complacency seems relevant in all phases of the Rubicon model. One can fail to overcome biases in deliberation, one can fail to commit, and one can fail with execution.

However, according to Doan's formulation, it seems that a commitment to taking sustainable action forms a prerequisite for lapsing into complacency. In terms of the Rubicon model, complacency, then, forms a danger in the executive phase only. If one has started on the way of taking sustainable action, through making a commitment, the motivational challenges of the executive phase cause the risk to be high that one will become complacent. Given that one's action soon stands out in surroundings where taking sustainable action is not commonplace, one has to overcome the motivational obstacles (see §6.6). Thus, there is a high risk that one becomes easily satisfied with one's achievements, which leads to one's commitment fading away. Again, my account accommodates the motivational issues Doan highlights.

Fourth, it could be suggested that morally corrupt agency can be explained in terms of simply failing to attend to morally important aspects of our lives. Kathie Jenni argues that, within a range of moral failings, the phenomenon of simple inattention has received little emphasis in the literature, which has concentrated mainly on self-deception (see below). Jenni formulates as follows:

> Humans' capacity to avoid unpleasant awareness is remarkable in its versatility. We use selective attention, creative reinterpretation, selective memory, and other strategies of self-deception to avoid confronting problematic aspects of our lives. We discredit sources of painful information, fail to follow up disturbing leads, selectively control the data we acquire. Most effectively of all, we simply let the momentum of the day-to-day pressures overwhelm moments of uneasiness or moral clarity that do occur.
>
> (Jenni 2003, 281)

From this formulation it seems that morally corrupt agents do indeed make use of inattention. Not paying attention to their motives for taking sustainable action is one of the ways through which they carry out their strategy. It is in the interest of morally corrupt agents that other agents should be inattentive to their resistance to act sustainably.

There is, though, one important reason why moral corruption cannot be classified under inattention. This is because of the wilful character of moral corruption and its deliberative and persistent nature. As they have been construed, morally corrupt agents may believe that they are moral in other parts of their lives than living up to Humanity's Challenge. It is only their commitment to sustainability that is nil. Not paying attention to the moral aspects of our lives, therefore, does not seem to be an adequate explanation of morally corrupt agency.

Morally corrupt agents, like other agents, hold certain beliefs about the world. Accordingly, the question could be raised whether their conduct necessarily involves holding certain beliefs that are mutually contradictory. Now recall that we are considering agents in a powerful position, who are strongly willed in the sense of being capable of effective agency in complex action situations. It is, then, interesting to investigate whether their morally corrupt conduct brings them into situations that contradict the beliefs that these agents hold when they are being effective. More precisely, we should answer the question of whether being a morally corrupt agent implies, according to the traditional definition of self-deception, that one intentionally believes some proposition, p, all the while knowing or believing truly ~p.

However, the traditional definition of self-deception seems too closely modelled on the concept of intentional *interpersonal* deception. That is, the paradoxes which apply to the traditional model do not occur when a person follows a deceitful strategy towards another person.[8] For an understanding of morally corrupt agency in relation to *self*-deception it seems that we need a broader concept of self-deception, one that allows interpreting 'to be deceived' to mean nothing more than 'to believe falsely' or 'to be mistaken about believing'. Although it may seem that this widening of 'to be deceived' leads to a conception of innocently being mistaken about a simple fact in the world, rather than one that involves self-deception, this can be remedied by specifying that the false belief is not accidental, but motivated by desire, anxiety, or some other emotion regarding p or related to p. We are thus looking for a concept of self-deception that includes an agent having a false belief p, being confronted with evidence that ~p, and having some desire or emotion that explains why p is believed and retained (Deweese-Boyd 2012, §3).

It could be objected here that self-deception so conceived does not differ much from wishful thinking. The only difference seems to be that the former is intentional whereas the latter is not. In reply, Ian Deweese-Boyd argues that self-deceivers recognise evidence against their self-deceptive belief, whereas wishful thinkers do not (Deweese-Boyd 2012, §3.1). Note that we have defined moral corruption as a decision agents have taken after they have agreed that collective action should be taken to mitigate climate change. The judgement that sustainable action is necessary has been based on the IPCC reports that are integrated in Humanity's Challenge. This implies that morally corrupt agents have accepted the evidence against the strategy they are pursuing. However, they have decided that the burden of doing something against climate change is not on them. Therefore, having construed moral corruption as a strategy of deliberately and persistently not making a commitment to take sustainable action, we can conclude that moral corruption is not a form of wishful thinking.

This leaves us with three questions that must be answered. First, what is it that morally corrupt agents falsely believe? Second, what evidence do they have that contradicts this belief? And, third, which desire explains their having the belief in question?

About the last question we can say that morally corrupt agents have the desire to preserve their convenient lives, both professionally and privately, not being bothered with the burden of doing something against the effects of climate change. They agree with everything that has been concluded about the normativity of our insights regarding the effects of human conduct on the climate, and they support the moral judgement, as laid down in the UNFCCC, that it is up to agents in the affluent countries to take a pro rata larger share of the burden than the developing countries. However, even whilst being aware of their own subscription to the norm of taking up Humanity's Challenge, morally corrupt agents desire that their personal and professional life remain unaffected. This means that, in terms of the Rubicon model, these agents may be able, in the deliberative phase, to select a sustainable action as the action (all-things-considered) that it is optimal to carry out. However, they do not commit to execute this action, which means that their action does not pass the commitment phase (see §5.2).

Note that this also answers the second question raised above about the evidence morally corrupt agents possess that contradicts their belief. Morally corrupt agents have accepted scientific evidence in the form of IPCC reports that, as a collective, people in the affluent countries must act to mitigate climate change. This means that their strategy cannot be universalised to people in the affluent countries, and the morally corrupt agent is aware that her course of conduct is incompatible with meeting Humanity's Challenge. In §9.4, however, I will argue that there is also evidence of a different kind that proves to be inconsistent with the agent's belief. I will argue that morally corrupt agency, in addition to being in contradiction to the normative principles of mitigating climate change, also has an effect on the agent's practical identity that forms an impediment to being an autonomous agent in the first place. That evidence, to be identified in §9.4, will be related to the effect on the agent's practical identity of morally corrupt agency and the holding of its underlying belief.

That is to say – and here I start to answer the first question – the way in which we can understand morally corrupt agency to be self-deceptive depends on how we formulate the content of the belief underlying this form of agency. As we have seen, the content of the belief that morally corrupt agents hold is that they can take a free ride on the efforts of others. These agents believe that the collective goal of mitigating climate change can be achieved without their making a contribution. It follows that the belief that morally corrupt agents entertain is that their decision to persistently not commit to taking sustainable action will ensure that their lives remain comfortable.

I will argue in §9.4, by referring to the constitution of our practical identity as agents, that this belief of morally corrupt agents is false. Perhaps in the short term the material situation of these agents can be considered convenient. When, however, we consider their practical identity, in the sense of being agents who are autonomous and efficacious in social cooperation with other agents, their lives will turn out to be deeply problematic for them. Their belief that they can continue with their lives conveniently, therefore, is indeed self-deceptive.

In particular, in §9.5, I argue that morally corrupt agents face difficulties with considering themselves as persons of moral integrity. Thereafter, in §9.6, I will argue that there are ways open to agents not to follow the self-deceptive strategy of moral corruption. This implies that moral corruption is under the control of the agent, who can reasonably be blamed for sticking to this strategy of not committing to take up Humanity's Challenge (Deweese-Boyd 2012, §5.1).

## 9.4 Moral corruption and our self-constitution as agents

We will now take up the task of investigating the effect of moral corruption on the agent's practical identity. As I stated in the previous section, the belief that morally corrupt agents entertain is that their decision to persistently not commit to meeting Humanity's Challenge will ensure that their lives remain comfortable. I argue that this belief on the part of morally corrupt agents is false. This conclusion follows from considering what moral corruption does to an agent's practical identity.

As I have construed them, morally corrupt agents are generally effective agents who are aware that having certain clear future-directed commitments is necessary for them to accomplish their strategies – if only in order to coordinate their activities with other agents. That is, on the one hand they form future-directed intentions with regard to the strategies they have chosen; and on the other hand, they do not make a commitment with regard to actions to take care of the distant future. Morally corrupt agents are aware that when they abandon prior intentions, they may be criticised for that on rational grounds. This is one of the reasons why they do not form intentions to act in favour of the environment: they could be accused of weakness of will if they do not manage to execute their plans. Morally corrupt agents do not make a commitment to care for the distant future by forming an intention or adopting a plan to do so, because this binds them to realise that intention or plan, which is not their objective. Being regarded by other agents as someone who fails at self-management is something morally corrupt agents particularly want to prevent.

Now we face the following questions: Can current people consider themselves to be moral agents – that is, assume to be acting morally – with regard to a certain class of actions (e.g. those affecting other current people or children or grandchildren), while at the same time acting defectively with regard to another class of actions, namely those concerned with meeting Humanity's Challenge? Which view on human agency would combine acting responsibly towards current people and our immediate offspring, and acting irresponsibly towards people living in the distant future?

In this book, I have investigated the motivational challenges that agents face when they take up Humanity's Challenge, by focusing on the agent's first-person perspective while she progresses through the deliberative, commitment, and executive Rubicon action phases. In order to address the motivational challenges, in a first step I have developed agential capabilities that agents need to acquire (see §7.2). In a second step, I have proposed a pathway for institutions that elaborates which interventions ought to be taken to develop these agential capabilities and facilitate sustainable action (see §7.3). However, working on Humanity's Challenge in this way may be useless, if people are to turn en masse to moral corruption through pursuing a strategy with the objective of maintaining their convenient status quo. Given the still-rising levels of global greenhouse gas emissions (see §1.5), we might, together with Stephen Gardiner, be tempted to conclude that the only outcome we can expect is an ethical tragedy, a scenario in which humanity will not be able to save itself, with the only hope being that the human race will be saved by causes external to it (Gardiner 2011a, 440).

There is, however, an overlooked resource that might lead pessimists to change their mind, once they see the ways in which it could be motivating for individuals. The resource I have in mind is reflection on how we ought to understand ourselves as moral agents and bringing this practical self-understanding into relation with the concept of moral corruption. As I will argue, this may lead to acknowledging that our practical identity becomes divided, if when we adhere to moral corruption. For this divided identity we are responsible, which means that we are severely hampered in considering ourselves candidates for possessing the virtue of integrity. However, the insight into our hampered integrity provides us with an indirect motive to reverse this evil, morally corrupt trait in ourselves. That is, we have a motive for taking up Humanity's Challenge from which we benefit immediately by means of the remedying of our divided identity, enabling us to remain candidates for the virtue of integrity.

In order to build up my argument, we start with emphasising the way in which we constitute ourselves through our agency. As already pointed out, we consider moral corruption as an action, as something that agents actually do, meaning that agents can corrupt themselves by acting in a morally corrupt way (see §9.3). For an assessment of the corrupting effect of moral corruption on our practical identity, we need to understand how our agency contributes to our self-constitution. Although a lot more can be said, we start with making three assumptions.

First, agents self-consciously make the choice as to which incentive, out of various incentives, they should act on, for reasons that are based on the potential working of the incentive upon them. This means that, for example, farmers make different choices regarding the working of incentives upon them than do police officers. Acting on an incentive in favour of a farmer's action constitutes one differently than acting on an incentive in favour of a police officer's action. Because farmers engage in courses of action that are specific to the role of farmer, just as police officers do in the context of their profession, farmers constitute their practical identities in a different way than police officers. Just as with these agents, we can see that people engaging in a course of sustainable action demonstrate that the self-conscious choice made to act on an incentive to contribute to meeting Humanity's Challenge outweighs the choice of a number of other incentives that do not favour such a course of action. One's choice to act upon an incentive that follows from one's commitment to take a sustainable course of action, therefore, constitutes one's practical identity in a different way than one's choice to follow incentives that do not favour sustainable courses of action.

Second, agents who engage in a course of action that extends over time realise that their practical identity is formed through their actions. These agents become aware that, after a while, the forming of their practical identity entails a kind of irreversibility. To continue with the above example, after having been a police officer for a number of years, it will become very difficult for one to take up the role of a professional farmer, because of the lasting effect of one's actions on the process of practical self-constitution. Moreover, agents become aware that they have constituted themselves, not only by the actions that they in a reflective stance morally approve of, but also by those actions that they disapprove of. More precisely, on the basis of the effect that the actions one morally disapproves of have on the formation of one's practical identity, one could form the intention of changing one's course of conduct in situations that are similar to those in which one has chosen a morally blameworthy action in the past.

Third, as discussed in §4.4, our concept of agency entails agents deliberating, planning, and executing actions for which they take responsibility. Within the process of practical self-constitution, agents are capable of integrating a sense of ownership of their actions with their aims, ideals, and aspirations. That is, to be a genuine moral agent, the effect of one's action on one's practical identity must be actively unified, must be integrated into the life of one narrative subject.

To summarise: we choose our actions on the basis of the potential working of incentives upon us; our actions make us who we are, which means that our practical identity is shaped over time; and we must integrate the effect of our actions on our practical identity into the self-told story of our life. We thus have an additional reason (that is, an *indirect* motive) to follow the right course of action, because if we are constantly doing the wrong thing we are not just failing at a first-order, direct level (which provides us with a *direct* motive to do the right action), but we are also perverting our practical identity in a way that compounds the scope of our wrongdoing.

On the basis of these preparatory considerations, we will now assess the effect of moral corruption on our practical identity. We will do this by considering an experienced and talented agent, A, being one and the same person, who is in a position to take up the roles and carry out the actions mentioned in the case descriptions in Chapter 3. As stated earlier, we consider A to be a powerful agent, who is strong-willed in the sense of being capable of effective agency in complex action situations. We now also suppose that A acts in a morally responsible way concerning those of her actions that affect current people, and in a morally corrupt way with respect to those of her actions that aim to meet Humanity's Challenge.

From the first assumption above, it follows that A chooses to act on incentives that make her contribute to ends that affect current people, whereas for ends that affect people in the distant future, A does not act on the respective incentives. More specifically, as a result of A's free riding as regards the ends that affect people in the distant future, A wills the end of mitigating climate change, but she does not will the means to those ends. Importantly, morally corrupt agents do not say that they do not will the ends of preserving the earth for future generations, but they keep questioning the means and fail to propose actions that better promote these ends than those actions on the table for deliberation and decision. As A judges in favour of taking environmentally friendly action, but does not make the commitment necessary to carry out that action, A acts according to instrumental coherence regarding actions that concern current people. However, in A's eyes, she can keep refusing to intend the means involved in actions that concern Humanity's Challenge (see §4.2).

When A deliberates about her actions and which ones to choose, being a capable and effective agent, she is conscious that her choices are hers; they are the product of her own activity. In particular, A views her choices regarding the ends promoted for current people and those for future people as an expression of her practical identity. However, she could be tempted to believe that actions promoting the ends for current people are constitutive of her practical identity, whereas actions in favour of people living in the distant future do not affect her practical identity, since she does not make a commitment to take those actions. To see what has gone wrong in A's reasoning, note that, on the basis of Kant's account that the agent's decision to follow evil must be considered an action (i.e. voluntary and purposive, see §9.3), it follows that A's practical identity is constituted by the actions that she carries out for the benefit of current people, and by her *action not to commit herself to act* with the objective of meeting Humanity's Challenge.

What effect then, does this distinction between these two sets of ends and corresponding actions have on A's practical identity? As we have seen throughout the conceptual-phenomenological analysis of Part II, the case studies of the "power grid rebels" and the purchase manager are paradigmatic for the characteristics of motivation for sustainable action. Due to the complexity of the plans these agents intend to implement, they must make public their plans, as they have set goals for themselves that inevitably affect the social context in which they operate. Without going public, these agents will not achieve much. The communicative action of making public their plans is a necessary condition for getting their social surroundings moving towards the end they want to achieve. The communication of their plans enables these agents to make use of the support of other agents, which they need for handling the complexity of executing their plans. The need to go public with their plans applies in particular to CEOs and politicians. As a consequence of their roles, these agents must make use of the practice of *delegation*, which means that others (who are by definition in a more subordinate position) are entrusted to further detail the plans, and subsequently execute them (Sunstein and Ullmann-Margalit, 1999).

Hence, to consider the influence that the two different sets of ends and corresponding actions, as related respectively to current people and people who will live in the distant future, have on A's practical identity, we shall concentrate on actions that A undertakes in a social context with other agents. Note that in this book we have assumed a minimally individualistic concept of institutions (see §2.2). Institutions are composed of human individuals, who are in an ongoing discourse with each other about the strategies, policies, and actions they are going to pursue as representatives in accordance with the goal of their institution. What the institution in fact represents is intersubjectively shared among the representatives. That is, we understand institutions not as legal entities, but as being the whole of the consciously coordinated activities of a group of agents. Moreover, agents must morally evaluate their different roles as against an integral set of normative standards. What agents do in their role as institutional agent and what they do as private individual, or in another institutional role, have to be integrated within their overall conception of how they understand themselves as a person.

We consider, therefore, A in her role as agent in an organisation. Because of her strategy to preserve her convenient status quo, A has deliberately chosen not to make public her reasons for not taking actions that contribute to mitigating climate change. This means that in the way A cooperates with other agents, she will share her reasons for actions that concern ends to be promoted for current people, whilst remaining silent about her reasons for not taking action in the interests of Humanity's Challenge.

Since A acts under complex contingent circumstances, she cannot control whether certain actions will be required of her and her organisation that are necessary for meeting Humanity's Challenge. Therefore, A's different treatment of the interests of current and future people leads to problems in the practical deliberation about common courses of action in her social interaction with other agents. This is the case because it is not clear a priori whether other agents will understand A leaving out certain types of argument from the discourse about these commonly shared courses of action.

In the course of their deliberation, it might appear to other agents, who do not distinguish between ends in favour of current people and ends in favour of future people the way A does – that is, who do not follow a strategy of moral corruption like A – that A is not sharing with them certain types of reasons, which could make it difficult for them to agree with A about what should be done in a situation in which certain actions are required. In common-sense language, other agents may refer to A as having a 'hidden agenda'. Following her choice of a morally corrupt course of conduct, A is not free to act according to what certain situations require of her. Given this, it follows that A will find it difficult to engage in relations of reciprocity with other agents, since they

> must be prepared to share their ends and reasons; to hold them jointly; and to act together. Reciprocity is the sharing of reasons, and you will enter into it only with someone you expect to deal with reasons in a rational way.
> (Korsgaard 1996a, 196)

This is what A fails to do.

As a side remark, one could argue that A could seek the company of other morally corrupt agents and remain on course with her strategy to keep up her convenient status quo. In effect, however, this would lead to a form of institutionalised moral corruption. These agents would become known by the public, either because of explicit communication by them, or because of people becoming aware of their anti-sustainability attitude as a collective. One way or another, their collective goal not to commit to take up Humanity's Challenge would bring them immediately into the camp of those suffering from Kant's third level of evil: malice or bad will. If A should act together with others *collectively* in a morally corrupt way this will likely have a disastrous effect on her practical identity as a result of the corrupting effect of A's behaviour on her moral character (see §9.3).

Hence, I return to the question about the effect of moral corruption on A's practical identity, assuming that she stays at Kant's second level of evil. Morally corrupt action, as I understand it, is an example of defective action in the sense that it fails "to constitute their agents as the unified authors of their actions" (Korsgaard 2009, 32). In fact, A's practical identity is constituted by A-as-adopting-ends-for-current-people and by A-as-not-committing-to-ends-for-future-people. These different ways of constituting her practical identity cause this identity to be divided, which raises the problem that A cannot unambiguously claim that her actions are issued from her constitution in that they come from a law she has given herself (Korsgaard 2009, 160). We can thus say that moral corruption causes a division of A's self into parts: one that is constituted by actions carried out to promote ends affecting current people, and another that is formed by the action not to make a commitment to take sustainable action. The contradiction that arises in A's practical identity is between the *moral* part that acts as-adopting-ends-for-current-people, and the other, *morally corrupt* part that acts as-not-committing-to-ends-for-future-people.

When making the further assumption that A performs a leadership role, the other agents with whom A socially interacts will be disoriented, not knowing at a given moment with which part of A's divided self they are dealing. Since A has a leadership role, as a result of which there will be strong reciprocal relations between A and the other agents, this uncertainty for the other agents negatively affects their motivation to act in accordance with any mutually agreed course of action in a given situation. For A and the other agents, it seems impossible to establish any unity between them – that is, to form a single common will. A's divided self hinders her and the other agents from meeting as noumena, that is from deliberating together and arriving at a shared decision (Korsgaard 2009, 190). As it seems fundamental for us as human beings that we understand ourselves as self-conscious agents, as noumena, *and* as participants in the social-communicative relations with other human beings, we might question A's ability to be an efficacious moral agent in social contexts at all.

What can we conclude, then, about the implications of moral corruption on an agent's practical identity? The least we can say about the self of a morally corrupt agent is that it is either entirely corrupt or divided. The wholeheartedly corrupt case we discussed above, hence I continue with the situation of a divided practical identity. On the one hand, this self is constituted by actions that pertain to the interests of current people, and on the other hand, it is formed by the agent's strategy of keeping her current, convenient status quo, which is being achieved by persistently not committing to take up Humanity's Challenge. There is thus an inherent contradiction in morally corrupt agency as we understand it in this book: these agents believe that they can preserve their convenient status quo, whereas through their conduct they acquire a divided practical identity, which renders their agency inefficacious in social contexts. Morally corrupt agency does not bring what the agent is expecting to achieve. On the contrary, such a strategy leads to a split identity that blocks the agent's capacity to act effectively in social cooperation with other agents who are not following a morally corrupt course of conduct.

Given the above argument, I now can reply to the two questions that remained open when I started to analyse moral corruption as self-deception (see §9.3). In addition to the evidence that morally corrupt agency is incompatible with meeting Humanity's Challenge, we can now add the insight that these agents must live with a divided practical identity. There is thus evidence that moral corruption is wrongful from the point of view of future generations, and there is also evidence that the effect of moral corruption on the agent renders her practical identity liable to be divided.

We, then, can answer the question about the belief that morally corrupted agents hold (that is, that they will be able to continue with their convenient status quo), if we take note of the second assumption made above. This assumption says that the forming of an agent's practical identity entails a kind of irreversibility. Recall that moral corruption means following a strategy. This means that one follows a course of non-sustainable action at least for a relevant period of time. It means also that one has identified with that course of action during that time, since one considers it an expression of one's practical identity. The irreversibility comes in when the agent realises that her strategy has formed her practical identity, which she cannot simply undo. The fact that she has willed her morally corrupt agency has shaped her practical identity until the moment that she has become aware of its effect on her.

The effect of moral corruption on their practical identity seems to consist in these agents being at war with themselves. The split parts of their identity fight for dominance over one another. Our analysis shows that the lives of these agents turn out to be deeply problematic for them. Morally corrupted agents cannot consider themselves autonomous and efficacious in social cooperation with other agents who do not follow a morally corrupt strategy. The schizophrenia of her practical identity, caused by moral corruption, effectively prevents the agent from achieving the end she has set for herself. Her belief that she can continue with her life conveniently, therefore, is indeed self-deceptive. Morally corrupt conduct rests on an illusion.

In the next section, I will discuss the question as to whether morally corrupt agents can see themselves as candidates for consideration as persons of moral integrity, and whether we can be held responsible for a divided self caused by morally corrupt agency.

## 9.5 Consequences of moral corruption for our integrity

The conclusion of the previous section leads directly to the question as to how our reflections about moral corruption, and the divided self that results from it, could be related to a conception of integrity. In this book we understand integrity to be a complex and thick virtue term, namely "a capacity to respond to change in one's values or circumstances, a kind of continual remaking of the self, as well as a capacity to balance competing commitments and values and to take responsibility for one's work and thought" (Cox *et al.* 2003, 41). Integrity so understood means that there certainly is a connection between living with integrity and living a morally good life. "Integrity is a complex aspect of character that serves to link or dissolve disparate goals, values, emotions, aspects of self and periods in one's life" (Cox *et al.* 2003, 56).

Can agents be morally corrupt, as we understand it in this book, and simultaneously consider themselves candidates for consideration as persons of integrity in the above sense? Note that I talk about being a 'candidate' for consideration as a person of integrity, since integrity seems hard to achieve. To balance one's competing commitments and values and to take responsibility for one's work and thought can be extremely difficult to accomplish depending on intrapersonal and other circumstances. It seems to me that we should be cautious about naming someone a person of integrity. Until the moment that we have achieved it (probably more towards the end of our lives), we remain modest candidates for possession of this virtue.

Recall that being morally corrupt means pursuing a strategy, with the objective of maintaining one's convenient status quo, which involves deliberately and persistently not making a commitment to take sustainable action, both in one's private and public roles. In other words, with respect to treating the interests of current people and future people, the least we can say about morally corrupt agents is that they operate with different moral standards. As a consequence, morally corrupt agents develop two sorts of self: one that cares for the interests of current people including those of the agent herself, and one that deliberately and persistently does not care for the interests of people who will live in the distant future (see §9.4).

Deliberately and persistently not forming those intentions that are necessary to overcome the motivational obstacles in taking action to meet Humanity's Challenge means that one of the selves of morally corrupt agents has the potential to act in good faith regarding current people, whereas their other self does not act in good faith with respect to people who will live in the far future. The morally corrupt agent does not seem to be able to form a moral point of view from which she can evaluate her actions regarding current people and future people on the basis of an integral set of normative standards. Since the capacity to evaluate her actions from an overall conception of how she understands herself as a person is a necessary condition for efficacious agency (see §2.2), it seems that the morally corrupt agent has problems with being a candidate for consideration as a person of integrity. For professional leaders in particular, in respect of whom we have legitimate expectations of consistency (e.g. politicians, CEOs), moral corruption seems to be a defeater of integrity (Cox *et al.* 2003, 112).

It seems, then, that an obvious question must be answered: Who is responsible for the effect of moral corruption on one's practical identity, leading to potential problems with one's integrity? In reply, I make use of the third assumption in §9.4. This assumption says that agents must integrate the effect of their actions on their practical identity into the self-told story of their life. This means that the agent is responsible for being a genuine moral agent, and has to cope with her divided practical identity in relation to being a candidate for consideration as a person of integrity.

In developing his account of moral integrity, Alasdair MacIntyre supports this view when he argues that moral agents can be justifiably held responsible in at least three respects. First, they are responsible for what in their actions is intentional. Second, they may be responsible for incidental aspects of their actions of which they should have been aware. Third, they may be responsible for at least some of the reasonably predictable effects of their actions (MacIntyre 1999, 312). Since morally corrupt agents have compartmentalised their commitments regarding current people and future people, it could be argued that these agents can be held responsible for their divided selves as well. Morally corrupt agents actively refuse to make a commitment that could lead to actions taken in the interest of people who will live in the distant future. These agents have deliberately closed their minds to certain possibilities for action which could be expected of them in both their individual and their institutional roles. As MacIntyre argues:

> This divided self has to be characterised by what it lacks. It is not only without any standpoint from which it can pass critical judgement on the standards governing its various roles, but it also must lack those virtues of integrity and constancy that are prerequisites for exercising the powers of moral agency.
>
> (MacIntyre 1999, 324)

I conclude that our integrity is at risk when we consistently do not care for the distant future whilst living an active life as a moral agent. Persistently not living up to our obligations to take up Humanity's Challenge negatively affects the circumstances in which future people will live, and it also casts doubts over us current people, when we still believe that we can be regarded as persons of integrity. Not striving for a sustainable world in the sense of being a morally corrupt agent – that is, not being able or willing to find good reasons and make the commitments needed for voluntarily carrying out actions to meet Humanity's Challenge – will hamper us in viewing ourselves as candidates for consideration as persons of integrity.

## 9.6 Two kinds of commitment to take up Humanity's Challenge

Up to this point, I have argued for two main consequences of following a morally corrupt course of conduct. First, the agent suffers from schizophrenia of her practical identity, which effectively hampers her autonomous and effective agency (see §9.4). Second, because of their strategy, morally corrupt agents cannot be viewed as candidates for consideration as persons of integrity (see §9.5). Moreover, they are morally responsible for their divided practical identity. On the basis of our insight into the consequences of moral corruption, I now propose ways in which we can desist from this form of defective agency. To this end, I argue for two sorts of commitment that we can make, which differ in degree.

The first question we discussed in arguing the reasons for starting the project of this book concerned the meta-ethical assumptions that are necessary to formulate the problem with motivation for sustainable action (see §1.3). In reply, I argued that we have *a* perhaps non-moral motive to take up Humanity's Challenge, which may be weak and typically overridden by other motives (see §1.4). Thereafter, when outlining the method of the book, I argued that the task of a philosophical understanding of the problem with motivation for sustainable action is to enhance the agent's practical reasoning about the question: What am I to do about taking up Humanity's Challenge? (see §2.4). Now, after having carried out the conceptual-phenomenological analysis of the agent's experience, and having analysed the concept of moral corruption, I claim that this investigation enables us to find an *indirect* motive for taking sustainable action.

As I have explained, not only will future generations benefit from sustainable action, but the current generation of agents benefits as well. Accepting and actively supporting policies for improving sustainable behaviour will be feasible if agents have not only internalised the moral motive to act on Humanity's Challenge, but also found another, indirect motive from which they benefit directly (see §2.4). Our case of acting on the motive to respect the rights of future generations shows that such a motive may motivate only weakly. By contrast, an *indirect* motive produces a value or good as a side effect from which the agent benefits.

Below, I formulate the indirect motive in relation to taking sustainable action. In this way, not only will future generations benefit from sustainable action; current agents will also benefit from the possibility to improve their lives in that they will be able to integrate *their own* interests – that is, in relation to themselves and their dear ones – with the end of meeting Humanity's Challenge. Mitigating climate change means that we commit to acting towards that end over decades and centuries (see §4.4). The indirect motive enables agents to integrate this commitment into their practical self-understanding, as it emerges from identifying with the narrative that is constituted by their actions, experiences, and psychological traits.

My argument for this first kind of commitment roughly takes the form of a *modus tollens*:

***Thesis***

1 *If I stick to morally corrupt behaviour, then I cannot consider myself an autonomous and efficacious agent capable of realising goals in social cooperation with other agents;*
2 *I do consider myself an autonomous and efficacious agent capable of realising goals in social cooperation with other agents;*
3 *Therefore, I ought not to adhere to morally corrupt behaviour, lest I should suffer from living with a divided practical identity for which I am morally responsible.*

As to the first premise, in this chapter, I have argued for the claim, concerning agents who are engaged in morally corrupt agency, that this form of agency is self-deceptive. Whereas morally corrupt agents believe that they can continue with their convenient lives without doing anything about Humanity's Challenge, I argued that they must face a kind of moral schizophrenia, which hampers their agency in social contexts (see §9.4). There is thus an inherent contradiction in morally corrupt agency: these agents believe that they can preserve their convenient status quo, whereas they suffer from a divided practical identity, which renders their agency inefficacious in social cooperation with other agents who do not follow a morally corrupt course of conduct. Therefore, the argument about the consequences of morally corrupt agency for her practical identity gives us the first premise that the agent cannot consider herself an autonomous and efficacious agent who is capable of realising goals in social cooperation with other agents.

The second premise follows from the understanding of the phenomenology of agency that consists of the subject of experiences (see §2.4) and the subject of agency (see §4.4), which we have elaborated. The assumption we made for choosing the conceptual-phenomenological method was that the agent has the experience of intentionality and freedom while she is acting, and that she is aware of this experience (see §2.4). Although there may be enormous constraints under which agents must operate, we assumed that the agent experiences that she can influence, at her own discretion, at least part of the course of events. Otherwise put, if we want to meet Humanity's Challenge, we need to understand ourselves as agents according to a conception of agency that enables us to view our actions across time in relation to our commitment. The conception of agency entails that agents are expected to be capable of following through with their action via the phases of the Rubicon model. Deliberation, commitment, execution, and evaluation are necessary in all roles and virtually all tasks that agents take up within social contexts. We thus necessarily attribute a certain form of autonomy to agents, and we take it that they are capable of accomplishing their tasks successfully.

I now conclude that, if we understand ourselves as autonomous and effective agents who consider themselves capable of realising goals in social cooperation with other agents, then we ought to desist from any morally corrupt conduct on our part, lest we have to live with grave schizophrenia, understood as a disunity of our practical identity for which we are morally responsible. Instead of deliberately and persistently not making a commitment to taking sustainable action, we ought to commit to taking sustainable action and commit to taking the measures that will be necessary for us to be sufficiently motivated to maintain our commitment. This does not imply that we can demand of ourselves that we act in a sustainable way all the time. Although we intend to act sustainably, due to the contingency of human life, many of our actions turn out to be suboptimally sustainable (see §1.2). However, we must make the commitment to take sustainable action. This is what we ought to do to take up Humanity's Challenge and fulfil our obligations to future generations.

By committing ourselves to act sustainably, we make use of the working of intentions upon us. Making a commitment arouses intentions to be formed when proposals for sustainable action pass the deliberative phase to the commitment phase. Although we may reconsider our intentions, intentions are conduct-controlling pro-attitudes that show inertia and function as inputs for further practical reasoning. If an agent's future-directed intention manages to survive until the time of action, and the agent notices that the time has arrived and nothing interferes, it will control the agent's action then. After forming a goal intention, agents start preparing for action execution. That is, they start to form their plans (see §5.2). Making a commitment, thereby forming an intention, gets us underway with sustainable action. For example, before I became aware of the problem of climate change, I may have taken the decision to buy a diesel-fuelled car. Our lives may be full of these non-sustainable, or – at least – suboptimal sustainable situations. However, now I am committed to taking up Humanity's Challenge, I can take the *resolution* either not to buy a next car, or to buy an electric one. When the time comes to replace my old car, the intention I have formed forces me to follow through, or to deliberately reconsider the intention and take a renewed decision.

Now that we are aware of the capabilities that are required for taking sustainable action (see §7.2), making the commitment to take up Humanity's

Challenge also means actively striving to acquire these capabilities. As it follows from one's autonomy as an agent, starting the learning process on one's own forms part of the commitment. In particular, the agent must engage with the institutions that have the task of facilitating the learning process. The agent must support the institutional policies and engage as a responsible citizen in debates about improving the measures that are (to be) implemented. Being committed to take sustainable action thus means that one accepts the individual responsibility to acquire the capabilities, and that one engages in designing and implementing the institutional policies (see §7.3).

I have thus given an account of what the first kind of commitment to take sustainable action entails. However, we may then ask: Does attempting to overcome one's tendency to decide in favour of moral corruption mean that we can consider ourselves candidates for possessing the virtue of integrity? Not quite. Achieving integrity requires more of us – if we understand this virtue as the capacity to respond to change in one's values or circumstances, a kind of continual remaking of the self, as well as a capacity to balance competing commitments and values and to take responsibility for one's work and thought (see §9.5). Being a candidate for consideration as a person of integrity requires that we make a second-order commitment to take up Humanity's Challenge and act on it.

As I have said before, making a second-order commitment to take up Humanity's Challenge entails making a commitment to meeting the motivational difficulties that one will face. The agent does not only commit to taking a sustainable action; she also commits to overcoming the motivational difficulties of which she has become aware. The second-order commitment goes further than the first kind of commitment in that it determines the agent's will. The second-order commitment ensures that the agent does everything that is required to make certain that she does not stop wanting to act sustainably and does not give up on her first-order commitment.

As we have seen in the case study of the bank manager (see Chapter 8), one's first-level commitment may still lead to situations of inter-institutional inaction or stalemate. The agent in such a case must change the contractual relationships with her stakeholders in order to resolve situations of stalemate, adopting a more or less 'whatever it takes' approach. In order to reach a new equilibrium in the relationships with her stakeholders, the agent must show a commitment to proceeding with implementing the sustainable policy that goes further than the first kind of commitment. This agent must convince her stakeholders, whilst maintaining a good and productive relationship with them, that these stakeholders must pay in part the price of the policy that promotes sustainable action. Her second-order commitment may enable the agent to cope with the significant risks that are likely to follow from her action, of which she perhaps was not aware when she committed to implementing the sustainable policy. Thus, not only the agent herself, but she *and* her network of stakeholders must change their mutually beneficial relationship with a view to acting upon Humanity's Challenge. That is, they must integrate their own interests with those of future generations and decide about what to do from the viewpoint of the continued existence of humanity.

An alternative formulation that comes close to what I mean by making the second-order commitment to take up Humanity's Challenge is that we impose upon ourselves a law that obliges us to use our "best endeavours" or "best efforts" (to use terms from UK and US contract law) to act sustainably. This entails that agents demand of themselves that they execute all reasonable courses of action required to meet the challenge. At the same time, the requirement that an agent use her best efforts means that the agent should give reasonable consideration to her own interests. That is, the agent must consider her own interests in order to be able to continue using her best endeavours to take sustainable action. The agent should not, for example, have to be prepared to run herself into financial ruin in order to carry out a sustainable action. A requirement of best endeavours, however, imposes an obligation to act in good faith and to the extent of the agent's own total capabilities.[9] Making the second-order commitment to take up Humanity's Challenge means that agents form those intentions needed to overcome the motivational obstacles that we found when we investigated the question "Why is it so hard to take sustainable action?" and use their best efforts to follow through.

As we hypothesised in the descriptive diagnosis of the commitment phase (see §5.6), agents who take sustainable action must somehow compensate for the (potential) impairment of their autonomy by making a stronger form of commitment. It seems to me that these agents derive this compensation from the ideal that they are contributing to meeting Humanity's Challenge. The fact that some of the agents were capable of overcoming the challenge to their autonomy may be grounded in their second-order commitment to take the sustainable action as described in the case studies. Thus, there seems to exist a kind of strengthening of one's autonomy, which does not particularly arise from the recognition in interaction with other agents, but which one obtains by freely and voluntarily deliberating on, committing to, and executing a sustainable action in light of the ideal of meeting Humanity's Challenge. Using her best efforts to achieve such an ideal renders the agent autonomous, enabling her to cope with a temporary degrading of her moral status by the social network in which she is acting.

Similarly, in the diagnosis of the executive phase (see §6.6) we observed that the paradigmatic agents can only rely on good physical and mental health as the source of their self-efficacy belief. Since these agents are able to manage their tasks until completion, it seems necessary to explain which source of motivation is available to them that enables them to overcome the motivational obstacles we have investigated. Now, by having an understanding of what it means to make a second-order commitment to taking sustainable action, I conjecture that this second-order commitment may have become a source of self-efficacy belief for these agents. Their second-order commitment enables them to perform self-control in that they resist reconsidering their intention. Their second-order commitment makes them form a habit that enables them to resist the temptations to which they are prone. No other person or institution holds these agents accountable for not carrying out their sustainable action. In my view, it is the second-order commitment of these agents that forms another source of the self-efficacy belief they need to overcome the motivational challenges they encounter.

After having proposed in Chapter 7 pathways for individual agents and institutions on the basis of the conceptual-phenomenological analysis of the phases of the Rubicon model, in this chapter I have gone on to develop the concept of moral corruption. I have argued why this form of defective agency is detrimental to us. In this section, I have elaborated two kinds of commitment that enable agents to stay away from moral corruption. Agents need to make the first kind of commitment to act in response to Humanity's Challenge. By doing this they remain autonomous and efficacious agents, who are capable of realising goals in social cooperation with other agents. In order to make the step to becoming persons of integrity, however, a second-order commitment is required. This commitment entails imposing upon oneself a law requiring the use of one's best efforts to take up Humanity's Challenge.

Therefore, from a methodological point of view the following can be concluded (see §2.4). From the normative requirement to meet Humanity's Challenge, I argued that we have a weak motive to take up this challenge by acting sustainably. From there, I investigated the motivational difficulties that agents face when they engage in sustainable action. This led to a descriptive diagnosis of the challenges, which I have given the status of a tentative hypothesis to be taken up by future empirical psychological research. At the same time, on the basis of the descriptive diagnosis, I was able to formulate pathways for individual agents and institutions (see Chapter 7). In addition, through analysing the concept of moral corruption, I have argued for an improved understanding of ourselves as agents. This enhanced practical self-understanding provides us with the indirect motive to prevent a moral schizophrenia of our practical identity by making the commitment to take up Humanity's Challenge, a commitment that must, in the face of severe inaction or stalemate affecting our proposed sustainable action, be extended to the level of a second-order commitment. Thus, our enriched practical self-understanding, like the normative requirement of meeting Humanity's Challenge, has normative force on us by providing the indirect motive for remaining unified moral agents and candidates for consideration as persons of moral integrity.

## Notes

1 In this chapter, I elaborate the argument of Mees (2015).
2 Note that Wood's summary of the predispositions is based on Kant (1974, 15–18).
3 See Di Paola (2013) for a discussion of Gardiner's parallel with Jane Austen's *Sense and Sensibility*.

4 This worldview is understood as entailing, amongst others: reliance on the market mechanism as primary regulatory principle; belief in unlimited economic growth; faith in abundance of resources; techno fix, i.e. technological solutions to social and environmental problems; self-interest as the principal motivator of behaviour.

5 See Miller (2011) and Feikema (2015, ch. 3) for an overview of the uses of corruption.

6 See Höffe (2014) for an elaborate view on the place of *Religion within the Boundaries of Mere Reason* within Kant's entire work.

7 The failure to form an intention is also the reason why moral corruption is not compatible with the first level of Kant's "radical evil", since at this level what matters is that one knows what one ought to do, that one intends to do it, but that one simply does not do it

8 These paradoxes are the following. The first one concerns the self-deceiver's state of mind. How can someone hold two contradictory beliefs at the same time? As the deceiver, she must believe ~p, and, as the deceived, she must believe p. But if it is not possible to be in a state of mind that involves believing self-consciously two contradictory propositions, then self-deception according to the traditional definition seems impossible. The second paradox is: How can someone intend to deceive herself – that is, follow a strategy of believing p whilst one is convinced that ~p? As the deceiver, she must be aware that she is deploying a deceitful strategy, whilst as the deceived she must not be aware of this strategy for it to be effective. This seems a difficult task to accomplish (Deweese-Boyd, 2012, §1).

9 I draw here on the literature concerning the meaning of "best endeavours" or "best efforts" in UK and US case law, e.g. Miller (2006). However, due to the scope of this book, I cannot go further into this topic.

# Concluding reflections

Now that we have reached the end of this book, we should raise the basic question: What have we learned about taking sustainable action and the motivational problems it involves? Obviously, I leave the final judgement to the reader. However, in response I would make the following observations.

At the start of the book, we considered a homeowner who lives in a middle-class town somewhere in an affluent country. She has an experience that affects her profoundly. Somehow, this experience forces her to change the view she has on her life. The refusal of the other homeowners to accept her proposal that they all place solar panels on their roofs leads to an unexpected awareness of what she actually cares about. Importantly, she finds out that what she cares about differs from what others care about, others who are significant to her. The situation of this homeowner comes to expression when, after an awkward and alienating pause in a neighbourhood gathering, one of the other homeowners utters the words "What's in it for me?" In this scenario, the other homeowners' response to her proposal (and to her) has a deep and lasting effect on her. After all, she had become convinced that accepting the offer for installation of the solar panels would be the appropriate thing to do, in order to live up to the challenge to do something about climate change.

The case study of this homeowner illustrates some of the factors that may affect an agent's motivation to take sustainable action. In five steps, I will now show how the Core Thesis of this book (see Introduction) may be of use in replying to the question "What's in it for me?"

First, the situation of the homeowner shows how bewildering situations can be in which agents need to agree on a common course of action, and some agents deem it self-evident that one should act in conformity with one's duties, whereas others ignore this entirely. In particular, it shows that we need reasonable assumptions, about which we ought to reach agreement, as a basis for discussing proposals and policies to take sustainable action. To this end, I have argued that we have sufficient grounds to accept the claims of climate science that the main causes of climate change are anthropogenic. Moreover, I have argued that there is a normative ethical basis for the claim that current people globally are under the moral obligation to substantially reduce their luxury greenhouse gas emissions. I have included these assumptions in a concise statement – Humanity's Challenge – which I argue it would be reasonable for agents globally to accept, though agents in the affluent countries ought to take it up first. To conclude, the groundwork of Part I leads to the first part of the Core Thesis: *We are concerned with agents in the affluent countries who have accepted the claim of climate science that the main causes of climate change are anthropogenic; these agents have also accepted that people in the affluent countries are under the moral obligation collectively to substantially reduce their luxury greenhouse gas emissions. This means that our focus is agents in the affluent countries who have accepted what we will call "Humanity's Challenge".*

Second, while acting sustainably, or acting in another morally relevant context, individual agents are confronted with motivational problems. In order to establish a structure for the investigation of how human motivation works in these action situations, we introduced the Rubicon model of action phases. Following the model, we investigated motivational problems that are characteristic for the phases of deliberation, commitment, and execution. The conceptual-phenomenological analysis of the Rubicon action phases we produced provides the descriptive diagnosis of the motivational problems with taking sustainable action. We do not claim that these problems are unique to agents taking sustainable action; however, they do indeed belong among the motivational features of sustainable action situations. As the descriptive diagnosis has been summarised at the end of each of Chapters 4 to 6 (see §4.5, §5.6, and §6.6) and in its entirety in §7.2, no further summarising is required here, except to say that Part II supports the claim: *While attempting to take up Humanity's Challenge by acting sustainably, individual agents are confronted with motivational problems in each of the action phases, be it during deliberation, when making a commitment, or while executing the action.*

Third, the diagnosis of the motivational difficulties for individual agents indicates that capabilities that agents could acquire in order to overcome the difficulties can be identified. These agential capabilities consist of three groups: capabilities of practical reasoning and judgement; capabilities of social coordination and recognition; and capabilities of the will. However, it follows from their responsibility as autonomous citizens in their constitutional democracies that individual agents not only *could* develop the agential capabilities, but *ought* to take these measures, or, at least, they ought to support the relevant institutions in helping them to acquire the capabilities. This is what the pathway for sustainable action for individual agents entails, which leads to the claim: *In order to overcome these motivational problems, individuals ought to take certain measures, …*

Fourth, the descriptive diagnosis provides a starting point for institutions to develop policies that facilitate sustainable behaviour. Armed with knowledge about the motivational difficulties that individual agents encounter, institutional agents can argue for policies that help people with overcoming precisely these difficulties. For example, we suggest institutional strategies for providing information to citizens, libertarian paternalistic strategies, and making use of moral education. Since knowledge of the motivational difficulties can be made public, and assuming that Humanity's Challenge is widely accepted, this enhances the probability that the people whose behaviour is the object of the policies will be the ones to approve the policies in affluent democratic societies. This is what the pathway for sustainable action for institutions entails, as summarised by the claim: ..., *supported by institutions that implement sustainable policies in constitutional democratic societies.*

However, both the individual and the institutional pathways may not be sufficient for meeting Humanity's Challenge. Agents may be required to make a second-order commitment to prevent situations of severe inaction or stalemate. The case study of the bank manager shows how making a second-order commitment could be viewed from the agent's perspective.

Fifth, what happens if we do not make a structural commitment to taking sustainable action, and persist in following a strategy on non-sustainable behaviour? Our account of the concept of moral corruption shows that we might find ourselves in a rather uncomfortable situation: even though we strive for the good life by doing our best for ourselves and significant others around us, we might end up suffering from moral schizophrenia due to our morally corrupt strategy, which affects future generations. Accordingly, apart from the moral motive to take up Humanity's Challenge, we have an indirect motive – that is, a motive that leads to a value or good as a side effect that obtains for the agent herself – for committing ourselves to taking sustainable action. Through staying away from moral corruption, we can prevent severe moral failure that has an impact on future generations. If we go a step further by making a second-order commitment to take up Humanity's Challenge, we might remain a candidate for possessing the virtue of integrity. This is what the pathway on sustainable action for humanity entails. Thus, the final claim of the Core Thesis is: *Finally, if we do not make a structural commitment to taking sustainable action, then we suffer as humanity, and as individual and institutional agents, from severe moral failure.*

Having concluded in this way, this means, finally, that the homeowner can now respond to the sceptic, who utters her reason for not being committed to taking sustainable action by asking "What's in it for me?" Raising this question reveals potentially many things about the questioner's motivation. It may show merely that the questioner needs a self-interested motive for taking any action. It may simply express a candid opinion that without self-interest no motivation is possible, which means that the questioner would not be interested in taking any action from which she herself does not immediately benefit. It may, however, also show a deep uncertainty on behalf of the questioner in that, although she accepts the normative background assumption of Humanity's Challenge, in a reflective stance she simply does not see herself succeeding with taking the action, unless she is motivated by a benefit that befalls her. Therefore, the homeowner's dialogue with the questioner needs to be sufficiently nuanced to take these potential motivations into account when she tries to respond to "What's in it for me?"

The starting point for the dialogue, obviously, should be the acceptance by both agents of Humanity's Challenge as a normative background assumption. From there, the agent could take the recalcitrant homeowner through the nuances of the motivational difficulties, as we have diagnosed them. She could offer help with finding ways to develop the agential capabilities. Or, being responsible citizens, these agents could join forces and initiate an institutional policy that facilitates private ownership of solar panels. Finally, the agent could elaborate the consequences of persisting in questioning the need for sustainable action by making "What's in it for me?" part of one's morally corrupt strategy. It might appeal to the recalcitrant homeowner that by committing to act sustainably she could benefit from the indirect motive of protecting herself from a divided practical identity. However, just as with Kant's "radical evil", the resisting homeowner's decision whether to end her morally corrupt behaviour or not will be voluntary and purposive. Hence, being motivated to take up Humanity's Challenge cannot be imposed on her, nor on anyone.

# References

## *Literature*


Abrahamse, W., Matthies, E. (2013), *Informational Strategies to Promote Pro-environmental Behaviour: Changing Knowledge, Awareness and Attitudes.* In: Steg, L., van den Berg, A., de Groot, J. (eds), *Environmental Psychology, an Introduction*, John Wiley & Sons, Chichester, pp. 223–232.

Althof, W., Berkowitz, M.W. (2013), *Character and Civic Education as a Source of Moral Motivation.* In: Heinrichs, K., Oser, F., Lovat, T. (eds), *Handbook of Moral Motivation: Theories, Models, Applications*, Sense Publishers, Rotterdam, pp. 567–583.

Anderson, J.H. (2014), *Autonomy and Vulnerability Entwined.* In: Mackenzie, C., Rogers, W., Dodds, S. (eds), *Vulnerability: New Essays in Ethics and Feminist Philosophy*, Oxford University Press, Oxford, pp. 134–161.

Anderson, J.H. (2016), *Structured Nonprocrastination: Scaffolding Efforts to Resist the Temptation to Reconstrue Unwarranted Delay.* In: Sirois, F.M., Pychyl, T.A. (eds), *Procrastination, Health, and Well-Being*, Elsevier, London, pp. 43–63.

Anderson, J.H., Honneth, A. (2005), *Autonomy, Vulnerability, Recognition, and Justice.* In: Christman, J., Anderson, J.H. (eds), *Autonomy and the Challenges of Liberalism: New Essays*, Cambridge University Press, Cambridge, UK, pp. 127–149.

Anscombe, G.E.M. (1957), *Intention*, Harvard University Press, Cambridge, Massachusetts.

Arnold, D.G. (ed.) (2011), *Global Climate Change*, Cambridge University Press, Cambridge, UK.

Asch, S. (1956), *Studies of Independence and Conformity: A Minority of One Against a Unanimous Majority*, Psychological Monographs, Vol. 70.

Atkinson, G., Dubourg, R., Hamilton, K., Munasinghe, M., Pearce, D., Young, C. (1997), *Measuring Sustainable Development*, Edward Elgar Publishing Ltd., Cheltenham.

Aven, T., Renn, O. (2009), *On Risk Defined as an Event Where the Outcome is Uncertain*, Journal of Risk Research, Vol. 12, No. 1, pp. 1–11.

Aydin, E. (2016), *Energy Conservation in the Residential Sector*, CentER, Center for Economic Research, University of Tilburg, The Netherlands.

Bandura, A. (1995), *Exercise of Personal and Collective Efficacy in Changing Societies.* In: Bandura, A. (ed.) (1995), *Self-Efficacy in Changing Societies*, Cambridge University Press, Cambridge, UK, pp. 1–45.

Batson, C.D., Thompson, E.R. (2001), *Why Don't Moral People Act Morally? Motivational Considerations*, Current Directions in Psychological Science, Vol. 10, No. 2, pp. 54–57.

Baumeister, R.F., Tierney, J. (2012), *Willpower: Why Self-Control Is the Secret to Success*, Penguin Books, London.

Baumgartner, C. (2005), *Umweltethik – Umwelthandeln. Ein Beitrag zur Lösung des Motivationsproblems*, Mentis, Paderborn.

Bayne, T. (2008), *The Phenomenology of Agency*, Philosophy Compass, Vol. 3, No. 1, pp. 182–202.

Bechtel, R., Churchman, A. (eds) (2002), *Handbook of Environmental Psychology*, John Wiley & Sons, Inc., New York.

Becker, C. (2006), *The Human Actor in Ecological Economics: Philosophical Approach and Research Perspectives*, Ecological Economics, Vol. 60, pp. 17–23.

Birnbacher, D. (2009), *What Motivates Us to Care for the (Distant) Future?* In: Gosseries, A., Meyer, L. (eds), *Intergenerational Justice*, Oxford University Press, Oxford, pp. 273–301.

Birnbacher, D. (2015), *Some Moral Pragmatics of Climate Change*. In: Birnbacher, D., Thorseth, M. (eds), *The Politics of Sustainability: Philosophical Perspectives*, Routledge, London, pp. 153–172.

Birnbacher, D. (2018), *Motivational Aspects of a Politics of Sustainability*. In: Düwell, M., Bos, G, and Van Steenbergen, N. (2018), *Towards the Ethics of a Green Future*, Routledge, London, pp. 154–166.

Birnbacher, D., Thorseth, M. (eds) (2015), *The Politics of Sustainability: Philosophical Perspectives*, Routledge, London.

Boatright, J. (2008), *Ethics in Finance*, 2nd edition, Blackwell Publishing, Malden, USA.

Bodanksy, D. (2016), *The Legal Character of the Paris Agreement*, Review of European, Comparative, and International Environmental Law, pp. 1–18.

Böhm, G., Tanner, C. (2013), *Environmental Risk Perception*. In: Steg, L., van den Berg, A., de Groot, J. (eds), *Environmental Psychology, an Introduction*, John Wiley & Sons, Chichester, pp. 15–26.

Bohman, J., Rehg, W. (2014), *Jürgen Habermas, The Stanford Encyclopedia of Philosophy*, Edward N. Zalta (ed.), https://plato.stanford.edu/entries/habermas/.

Bolderdijk, J.W., Steg, L., Geller, E.S., Lehman, P.K., Postmes, T. (2012), *Comparing the Effectiveness of Monetary versus Moral Motives in Environmental Campaigning*, Nature Climate Change, www.nature.com/natureclimatechange.

Bookchin, M. (1990), *The Philosophy of Social Ecology*, Black Rose, Montreal.

Booth, C. (2009), *A Motivational Turn for Environmental Ethics*, Ethics & The Environment, Vol. 14, No. 1, pp. 53–78.

Booth, C. (2012), *Bystanding and Climate Change*, Environmental Values, Vol. 21, No. 4, pp. 397–416.

Bos, G., Düwell, M. (eds) (2016), *Human Rights and Sustainability*, Routledge, London.

Bourban, M. (2014), *Climate Change, Human Rights and the Problem of Motivation*, De Ethica, Vol. 1, No. 1, pp. 37–52.

Bratman, M.E. (1989), *Intention and Personal Policies*, Philosophical Perspectives, Vol. 3, Philosophy of Mind and Action Theory, pp. 443–469.

Bratman, M.E. (1999), *Intention, Plans and Practical Reason*, CSLI Publications, USA.

Bray, D. (2010), *The Scientific Consensus of Climate Change Revisited*, Environmental Science & Policy Vol. 13, pp. 340–350.

Brennan, A., Lo, Y-S. (2015), *Environmental Ethics, The Stanford Encyclopedia of Philosophy*, Edward N. Zalta (ed.), http://plato.stanford.edu/entries/ethics-environmental/.

Brink, D.O. (1997), *Moral Motivation*, Ethics, Vol. 108, No. 1, pp. 4–32.

Brister, E. (2013), *Global Warming and the Problem of Failed Intentions*, Philosophy and Public Issues, Vol. 3, No. 1, pp. 247–271.

Buss, S. (2014), *Personal Autonomy*, The Stanford Encyclopedia of Philosophy, Edward N. Zalta (ed.), http://plato.stanford.edu/entries/personal-autonomy/.

Calhoun, C. (1995), *Standing for Something*, The Journal of Philosophy Vol. 92, pp. 235–260.

Campbell (2000), *Diagnosing Agency*, Philosophy, Psychiatry & Psychology, Vol. 7, No. 2, pp. 107–119.

Caney, S. (2010), *Climate Change, Human Rights, and Moral Thresholds*. In: Gardiner, S.M., Caney, S., Jamieson, D., Shue, H. (eds), *Climate Ethics, Essential Readings*, Oxford University Press, New York, pp. 163–177.

Caney, S. (2014), *Climate Change, Intergenerational Equity and the Social Discount Rate*, Politics, Philosophy & Economics, Vol. 13, No. 4, pp. 320–342.

Care, N. (1982), *Future Generations, Public Policy, and the Motivation Problem*, Environmental Ethics, Vol. 4, pp. 195–213.

Cox, D., La Caze, M., Levine, M. (2003), *Integrity and the Fragile Self*, Ashgate Publishing, Hants, UK.

Christman, J. (2015), *Autonomy in Moral and Political Philosophy, The Stanford Encyclopedia of Philosophy*, Edward N. Zalta (ed.), http://plato.stanford.edu/entries/autonomy-moral/.

Coeckelberg, M. (2016), *Environmental Skill*, Routledge, London.

Cook, J., Nuccitelli, D., Green, S., Richardson, M., Winkler, B., Painting, R., Way, R., Jacobs, P., Skuce, A. (2013), *Quantifying the Consensus on Anthropogenic Global Warming in the Scientific Literature*, Environmental Research Letters Vol. 8, 024024.

Cook, J., Oreskes, N., Doran, P.T., Anderegg, W.R.L., Verhaggen, B., Maibach, E.W., Carlton, J.S., Lewandowsky, S., Skuce, A.G., Green, S.A., Nuccitelli, D., Jacobs, P., Richardson, M., Winkler, B., Painting, R., Rice, K. (2016), *Consensus on Consensus: A Synthesis of Consensus Estimates on Human-Caused Global Warming*, Environmental Research Letters Vol. 11, 048002.

CRD II (2006), *Directive 2006/48/EC of the European Parliament and of the Council*, June 2006.

CRD IV (2013), *Directive 2013/36/EU of the European Parliament and of the Council*, June 2013.

Cripps, E., (2013), *Climate Change and the Moral Agent: Individual Duties in an Interdependent World*, Oxford University Press, Oxford.

Dancy, J. (2000), *Practical Reality*, Oxford University Press, Oxford.

Davidson, D. (1963), *Actions, Reasons, Causes*, The Journal of Philosophy, Vol. 60, No. 23, pp. 685–700.

Davidson, D. (2001), *Essays on Actions and Events*, Oxford University Press, Oxford.

Degenhardt, L. (2002), *Why Do People Act in Sustainable Ways? Results of an Empirical Study of Lifestyle Pioneers*. In: Schmuck, P., Schultz, W. (eds), *Psychology of Sustainable Development*, Springer, New York, pp. 123–148.

Deweese-Boyd, I. (2012), *Self-Deception, The Stanford Encyclopedia of Philosophy*, Edward N. Zalta (ed.), http://plato.stanford.edu/entries/self-deception/.

Dietrich, J. (2002), *Moralpedagogik*. In: Düwell, M., Hübenthal, Chr., Werner, M. (eds), *Handbuch Ethik*, Verlag J.B. Metzler, Stuttgart, pp. 423–28.

Di Paola, M. (2013), *Climate Change and Moral Corruption*, Philosophy and Public Issues, Vol. 3, No. 1, pp. 55–67.

Di Paola, M. (2014), *Climate Change: Who Does What, Why and How*. In: Di Paola, M., Pellegrino, G. (eds), *Canned Heat, Ethics and Politics of Global Climate Change*, Routledge, London, pp. 144–159.

Di Paola, M., Pellegrino, G. (eds) (2014), *Canned Heat, Ethics and Politics of Global Climate Change*, Routledge, London.

DNB (2016), Schotten, G., Van Ewijk, S., Regelink, M., Dicou, D., Kakes, J., *Tijd voor transitie: Een verkenning van de overgang naar een klimaatneutrale economie*, Occasional Study: de Nederlandsche Bank, Amsterdam.

Doan, M. (2013), *Climate Change and Complacency*, Hypathia, Vol. 10, No. 10, pp. 1–17.

Doran, P., Kendall Zimmerman, M. (2009), *Examining the Scientific Consensus on Climate Change*, Eos, Vol. 90, No. 3, pp. 22–23.

Dunlap, R.E., Van Liere, K.D. (1978), *The "New Environmental Paradigm"*, Journal of Environmental Education, Vol. 9, pp. 10–19.

Dunlap, R.E., McCright, A.M. (2008), *A Widening Gap: Views on Climate Change*, Environment: Science and Policy for Sustainable Development, Vol. 50, No. 5, pp. 26–35.

Düwell, M. (2014), *Bioethics: Domains, Theories, Cases*, Routledge, London.

Düwell, M., Hübenthal, C., Werner, M. (eds) (2002), *Handbuch Ethik*, Verlag J.B. Metzler, Stuttgart.

Düwell, M., Bos, G, and Van Steenbergen, N. (2018), *Towards the Ethics of a Green Future*, Routledge, London.

Farbotko, C., Lazrus, H. (2012), *The First Climate Refugees? Contesting Global Narratives of Climate Change in Tuvalu*, Global Environmental Change, Vol. 22, No. 2, pp. 382–390.

Feikema, L. (2015), *Still Not at Ease*, Department of Philosophy and Religious Studies, Utrecht University, Quaestiones Infinitae no. 85.

Frankfurt, H. (1971), *Freedom of the Will and the Concept of the Person*, The Journal of Philosophy, Vol. 68, No. 1, pp. 5–20.

Fransson, N., Gärling, T. (1999), *Environmental Concern: Conceptual Definitions, Measurement Methods, and Research Findings*, Journal of Environmental Psychology Vol. 19, pp. 369–382.

Gallagher, S. (2000), *Philosophical Conceptions of the Self: Implications for Cognitive Science*, Trends in Cognitive Science, Vol. 4, No. 1, pp. 14–21.

Gallagher, S., Zahavi, D. (2015), *Phenomenological Approaches to Self-Consciousness, The Stanford Encyclopedia of Philosophy*, Edward N. Zalta (ed.), http://plato.stanford.edu/entries/self-consciousness-phenomenological/.

Gardiner, S.M. (2006), *A Perfect Moral Storm: Climate Change, Intergenerational Ethics and the Problem of Moral Corruption*, Environmental Values Vol. 15, pp. 397–413.

Gardiner, S.M. (2011a), *A Perfect Moral Storm: The Ethical Tragedy of Climate Change*, Oxford University Press, New York.

Gardiner, S.M. (2011b), *Is No One Responsible for Global Environmental Tragedy? Climate Change as a Challenge to Our Ethical Concepts*. In: Arnold, D.G. (ed.)

(2011), *Global Climate Change*, Cambridge University Press, Cambridge, UK, pp. 38–59.
Gardiner, S.M., Caney, S., Jamieson, D., Shue, H. (eds) (2010), *Climate Ethics, Essential Readings*, Oxford University Press, New York.
Garvey, J. (2008), *The Ethics of Climate Change. Right and Wrong in a Warming World*, Continuum, London.
Gifford, R. (2007), *Environmental Psychology and Sustainable Development: Expansion, Maturation, and Challenges*, Journal of Social Issues, Vol. 63, No. 1, pp. 199–212.
Gifford, R. (2008), *Psychology's Essential Role in Alleviating the Impacts of Climate Change*, Canadian Psychology, Vol. 49, No. 4, pp. 273–280.
Gifford, R. (2011), *Dragons of Inaction, Psychological Barriers That Limit Climate Change Mitigation and Adaptation*, American Psychologist, Vol. 66, No. 4, pp. 290–302.
Gigengack, D. (2013), *Activisten op de werkvloer*. In: NRC Handelsblad/The Optimist, 29 June 2013.
Gledhill, J. (2014), *The Basic Structure of the Institutional Imagination*, Journal of Social Philosophy, Vol. 45, No. 2, pp. 270–290.
Gollwitzer, P.M. (1990), *Action Phases and Mind-Sets*. In: Higgins, E., Sorrentino, R. (eds), *Handbook of Motivation and Cognition*, Vol. 2, The Guilford Press, New York, pp. 53–91.
Gollwitzer, P.M. (1999), *Implementation Intentions: Strong Effects of Simple Plans*, The American Psychologist, Vol. 54, No. 7, pp. 493–503.
Goodall, C. (2011), *How To Live a Low-Carbon Life*, 2nd edition, Earthscan, London.
Gosseries, A., Meyer, L. (eds) (2009), *Intergenerational Justice*, Oxford University Press, Oxford.
Gottlieb, P. (2006), *The Practical Syllogism*. In: Kraut, R. (ed.) (2006), *The Blackwell Guide to Aristotle's Nicomachean Ethics*, Wiley Online Library, pp. 218–233.
Graumann, C. (2002), *The Phenomenological Approach to People-Environment Studies*. In: Bechtel and Churchman (eds), *Handbook of Environmental Psychology*, John Wiley & Sons, Inc., New York, pp. 95–113.
Halbig, C. (2002), *Anerkennung*. In: Düwell, M., Hübenthal, C., Werner, M. (eds), *Handbuch Ethik*, Verlag J.B. Metzler, Stuttgart, pp. 297–301.
Hardin, R. (2003), *The Free Rider Problem, The Stanford Encyclopedia of Philosophy*, Edward N. Zalta (ed.), http://plato.stanford.edu/entries/free-rider/.
Harnacke, C. (2015), *From Human Nature to Moral Judgments*, Department of Philosophy and Religious Studies, Utrecht University, Quaestiones Infinitae no. 88.
Heckhausen, H., Gollwitzer, P.M., Weinert, F.E. (eds) (1987), *Jenseits des Rubicon: Der Wille in den Humanwissenschaften*, Springer Verlag, Berlin.
Heckhausen, J., Heckhausen, H. (eds) (2010), *Motivation and Action*, Cambridge University Press, Cambridge, UK.
Heidbrink, L. (2015), *Libertarian Paternalism, Sustainable Self-Binding and Bounded Freedom*. In: Birnbacher and Thorseth (eds), *The Politics of Sustainability: Philosophical Perspectives*, Routledge, London, pp. 173–194.
Heinrichs, K. (2013), *Moral Motivation in the Light of Action Theory: Perspectives on Theoretical and Empirical Progress*. In: Heinrichs, K., Oser, F., Lovat, T. (eds), *Handbook of Moral Motivation: Theories, Models, Applications*, Sense Publishers, Rotterdam, pp. 623–657.

Heinrichs, K., Oser, F., Lovat, T. (eds) (2013), *Handbook of Moral Motivation: Theories, Models, Applications*, Sense Publishers, Rotterdam.

Held, D. (2014), *Climate Change, Global Governance and Democracy: Some Questions.* In: Di Paola, M., Pellegrino, G. (eds), *Canned Heat, Ethics and Politics of Global Climate Change*, Routledge, London, pp. 17–28.

Hiller, A. (2011), *Climate Change and Individual Responsibility*, The Monist, Vol. 94, No. 3, pp. 349–368.

Höffe, O. (2014), *Holy Scriptures within the Bounds of Mere Reason: Kant's Reflections.* In: Michalson, G. (ed.), *Kant's Religion within the Boundaries of Mere Reason*, Cambridge University Press, Cambridge, UK, pp. 10–30.

Holton, R. (2009), *Willing, Wanting, Waiting*, Clarendon Press, Oxford.

Honneth, A. (1995), *The Struggle for Recognition*, translated by J.H. Anderson, Polity Press, Cambridge, UK.

Hourdequin, M. (2010), *Climate, Collective Action and Individual Ethical Obligations*, Environmental Values, Vol. 19, pp. 443–464.

Hübenthal, C. (2007), *Maxime.* In: Becker, M., Stokkom, B. van, Tongeren, P. van, Wills, J.-P., *Lexicon van de Ethiek*, Van Gorcum, Assen, The Netherlands, pp. 201–203.

Hulme, M., Mahony, M. (2010), *Climate Change: What Do We Know about the IPCC?* Progress in Physical Geography Vol. 34 No. 5, pp. 705–718.

Inzlicht, M., Schmeichel, B.J. (2012), *What Is Ego Depletion: Toward a Mechanistic Revision of the Resource Model of Self-Control*, Perspectives on Psychological Science, Vol. 7 No.5, pp. 450–463.

IPCC (1990), Houghton, J.T., Jenkins, G.J., Ephraums, J.J. (eds), *Climate Change: The IPCC Scientific Assessment*, Cambridge University Press, Cambridge, UK.

IPCC (1992), Houghton, J.T., Callander, B.A., Varney, S.K. (eds), *Climate Change 1992: The Supplementary Report to the IPCC Scientific Assessment*, Cambridge University Press, Cambridge, UK.

IPCC (1995), Houghton, J.T., Meira Filho, L.G., Callander, B.A., Harris, N., Kattenberg, A., Maskell, K. (eds), *Climate Change 1995, The Science of Climate Change*, Cambridge University Press, Cambridge, UK.

IPCC (2001), Houghton, J.T., Ding, Y., Griggs, D.J., Noguer, M., Van der Linden, P.J., Dai, X., Maskell, K., and Johnson, C.A. (eds), *Climate Change 2001: The Scientific Basis*, Cambridge University Press, Cambridge, UK.

IPCC (2007), *Summary for Policymakers*, Climate Change 2007: Synthesis Report.

IPCC (2014a), *Approved Summary for Policymakers*, Climate Change 2014: Synthesis Report.

IPCC (2014b), Edenhofer, O., Pichs-Madruga, R., Sokona, Y., Farahani, E., Kadner, S., Seyboth, K., Adler, A., Baum, I., Brunner, S., Eickemeier, P., Kriemann, B., Savolainen, J., Schlömer, S., von Stechow, C., Zwickel T., and Minx, J.C. (eds), *Summary for Policymakers.* In: *Climate Change 2014, Mitigation of Climate Change*, Cambridge University Press, Cambridge, UK.

IPCC (2018), Masson-Delmotte, V., Zhai, P., Pörtner, H.O., Roberts, D., Skea, J., Shukla, P.R., Pirani, A., Moufouma-Okia, W., Péan, C., Pidcock, R., Connors, S., Matthews, J.B.R., Chen, Y., Zhou, X., Gomis, M.I., Lonnoy, E., Maycock, T., Tignor, M., Waterfield, T. (eds), *Summary for Policymakers.* In: *Global warming of 1.5°C. An IPCC Special Report on the impacts of global warming of 1.5°C above pre-industrial levels and related global greenhouse gas emission pathways, in the context of strengthening the global response to the threat of climate change, sustainable development, and efforts to eradicate poverty*, World Meteorological Organization, Geneva.

Jamieson, D. (2002), *Morality's Progress: Essays on Humans, Other Animals, and the Rest of Nature*, Clarendon Press, Oxford.

Jamieson, D. (2010), *Climate Change, Responsibility, and Justice*, Social and Engineering Ethics, Vol. 16, pp. 431–445.

Jamieson, D. (2017), *Reason in a Dark Time*, Oxford University Press, Oxford.

Janzing, B. (2002), *Baden unter Strom*, Dold Verlag, Germany.

Jenni, K. (2003), *Vices of Inattention*, Journal of Applied Philosophy, Vol. 20, No. 3, pp. 279–295.

Jensen, M., Meckling, W. (1976), *Theory of the Firm: Managerial Behaviour, Agency Costs and Ownership Structure*, Journal of Financial Economics Vol. 3, pp. 305–360.

Kagan, S. (1989), *The Limits of Morality*, Clarendon Press, Oxford.

Kalis, A. (2009), *Failures of Agency*, Department of Philosophy and Religious Studies, Utrecht University, Quaestiones Infinitae no. 57.

Kant, I. (1974), *Die Religion innerhalb der Grenzen der bloßen Vernunft*, Reclam, Stuttgart.

Kant, I. (2002), *Groundwork for the Metaphysics of Morals*, translated by A.W. Wood, Yale University Press, New Haven.

Kim, M., Surroca, J., Tribó, J.A. (2014), *Impact of Ethical Behaviour on Syndicated Loan Rates*, Journal of Banking and Finance, Vol. 38, pp. 122–144.

Kleingeld, P. (2007), *Kant: de menselijke hang naar het kwaad*. In: Kinneging, A., Wiche, R. (2007), *Van Kwaad tot Erger*, Spectrum, Utrecht, pp. 227–241.

Klöckner, C. (2013), *How Powerful Are Moral Motivations in Environmental Protection? An Integrated Model Framework*. In: Heinrichs, K., Oser, F., Lovat, T. (eds), *Handbook of Moral Motivation: Theories, Models, Applications*, Sense Publishers, Rotterdam, pp. 447–472.

Kockelmans, J. (2006), *Phenomenology*. In: Audi, R., *The Cambridge Dictionary of Philosophy*, 2nd edition, Cambridge University Press, New York, pp. 664–666.

Koger, S.M., Winter, D.D.N. (2010), *The Psychology of Environmental Problems*, 3rd edition, Psychology Press, New York.

Koller, T., Goedhart, M., Wessels, D. (2010), *Valuation: Measuring and Managing the Value of Companies*, 5th edition, John Wiley & Sons, New Jersey.

Kollmus, A., Agyeman, J. (2002), *Mind the Gap: Why Do People Act Environmentally and What Are the Barriers to Pro-Environmental Behaviour?* Environmental Education Research, Vol. 8, No. 3, pp. 239–260.

Kolodny, N., Brunero, J. (2013), *Instrumental Rationality, The Stanford Encyclopedia of Philosophy*, Edward N. Zalta (ed.), https://plato.stanford.edu/entries/rationality-instrumental/.

Korsgaard, C. (1996a), *Creating the Kingdom of Ends*, Cambridge University Press, Cambridge, UK.

Korsgaard, C. (1996b), *The Unity of Moral Agency: A Kantian Response to Parfit*. In: Korsgaard, C. (1996a), *Creating the Kingdom of Ends*, Cambridge University Press, Cambridge, UK, pp. 363–397.

Korsgaard, C. (2009), *Self-Constitution. Agency, Identity and Integrity*, Oxford University Press, New York.

Korta, K., Perry, J. (2015), *Pragmatics, The Stanford Encyclopedia of Philosophy*, Edward N. Zalta (ed.), http://plato.stanford.edu/entries/pragmatics/.

Koslowski, P. (2012), *The Ethics of Banking*, Springer, Dordrecht.

Kretz, L., (2012), *Climate Change: Bridging the Theory-Action Gap*, Ethics and the Environment, Vol. 17, No. 2, pp. 9–27.

LMA (2016), *20 Years in the Loan Market*, Loan Market Association, London.

LMA (2019), *Sustainability Linked Loan Principles*, Loan Market Association, London.

Luhmann, N. (1977), *Differentiation of Society*, The Canadian Journal of Sociology, Vol. 2, No. 1, pp. 29–53.

MacIntyre, A. (1999), *Social Structures and Their Threats to Moral Agency*, Philosophy Vol. 74, pp. 311–329.

Makkreel, R. (2015), *Orientation and Judgement in Hermeneutics*, University of Chicago Press, Chicago.

McConnell, T. (2014), *Moral Dilemmas, The Stanford Encyclopedia of Philosophy*, Edward N. Zalta (ed.), http://plato.stanford.edu/entries/moral-dilemmas/.

McIntyre, A. (2006), *What Is wrong with Weakness of Will?* The Journal of Philosophy, Vol. 103, No. 6, pp. 284–311.

McKinnon, C. (2011), *Climate Change Justice: Getting Motivated in the Last Chance Saloon*, Critical Review of International Social and Political Philosophy, Vol. 14, No. 2, pp. 195–213.

McKinnon, C. (2012), *Climate Change and Future Justice*, Routledge, London.

McQueen, A. (2016), *Political Realism and Moral Corruption*, European Journal of Political Theory, August 2016.

Mead, G.H. (1934), *Mind, Self and Society from the Standpoint of a Social Behaviorist*, University of Chicago Press, Chicago.

Mees, R. (2010), *Intergenerational Justice and Personal Identity*, MA thesis, Department of Philosophy and Religious Studies, Utrecht University.

Mees, R. (2015), *Sustainable Action and Moral Corruption*. In: Birnbacher, D., Thorseth, M. (eds), *The Politics of Sustainability: Philosophical Perspectives*, Routledge, London, pp. 109–126.

Mekonnen, M.M., Hoekstra, A.Y. (2010), *The Green, Blue and Grey Water Footprint of Farm Animals and Animal Products*, Value of Water Research Report Series No. 48, UNESCO-IHE Institute for Water Education, Delft.

Mele, A. (2002), *Autonomy, Self-Control, and Weakness of Will*. In: Kane, R. (2002), *The Oxford Handbook of Free Will*, Oxford University Press, Oxford, pp. 529–548.

Milgram, S. (1974), *Obedience to Authority: An Experimental View*, Harper & Row, New York.

Millennium Ecosystem Assessment (2005), *Ecosystems and Human Well-being: Synthesis*, Island Press, Washington, DC.

Miller, S. (2011), *Corruption, The Stanford Encyclopedia of Philosophy*, Edward N. Zalta (ed.), http://plato.stanford.edu/entries/corruption/.

Miller, S. (2012), *Social Institutions, The Stanford Encyclopedia of Philosophy*, Edward N. Zalta (ed.), http://plato.stanford.edu/entries/social-institutions/.

Miller, Z. (2006), *Best efforts? Differing Judicial Interpretations of a Familiar Term*, Arizona Law Review Vol. 48, pp. 615–638.

Monin, B., Merrit, A. (2012), *Moral Hypocrisy, Moral Inconsistency, and the Struggle for Moral Integrity*. In: Mikulincer, M., Shaver, P.R., *The Social Psychology of Morality: Exploring the Causes of Good and Evil*, American Psychological Association.

Morgan-Knapp, C., Goodman, C. (2014), *Consequentialism, Climate Harm and Individual Obligations*, Ethical Theory and Moral Practice (2015) Vol. 18, pp. 177–190.

Murphy, L.B. (1999), *Institutions and the Demands of Justice*, Philosophy and Public Affairs Vol. 27, No. 4, pp. 251–291.

Næss, A. (1973), *The Shallow and the Deep, Long-Range Ecology Movement*, Inquiry Vol. 16, pp. 151–155.

Nihlén Fahlquist, J. (2009), *Moral Responsibility for Environmental Problems: Individual or Institutional?* Journal of Agricultural Environmental Ethics, Vol. 22, pp. 109–124.

Norton, B., Hutchins, M., Stevens, E., Maple, T. (eds) (1995), *Ethics on the Ark*, Smithsonian Institution Press, Washington, DC.

Olivier, J., Janssens-Maenhout, G., Peters, J. (2012), *Trends in Global $CO_2$ Emissions*, PBL Netherlands Environmental Assessment Agency.

O'Neill, J. (1992), *The Varieties of Intrinsic Value*, Monist Vol. 75, pp. 119–137.

Oreskes, N. (2004), *Beyond the Ivory Tower: The Scientific Consensus on Climate Change*, Science, Vol. 306, p. 1686.

Oser, F. (2013), *Models of Moral Motivation*. In: Heinrichs, K., Oser, F., Lovat, T. (eds), *Handbook of Moral Motivation: Theories, Models, Applications*, Sense Publishers, Rotterdam, pp. 7–24.

Ott, K. (1996), *Vom Begründen zum Handeln: Aufsätze zur angewandten Ethik*, Attempto Verlag GmbH, Tübingen.

Ott, K. (2004), *Essential Components of Future Ethics*. In: Döring, R., Rühs, M. (eds), *Ökonomische Rationalität und Praktische Vernunft*, Königshausen & Neumann, Würzburg, pp. 83–108.

Ott, K. and Döring, R. (2008), *Theorie und Praxis Starker Nachhaltigkeit*, 2nd edition, Metropolis-Verlag, Marburg.

Pacherie, E. (2008), *The Phenomenology of Action: A Conceptual Framework*, Cognition, Vol. 107, No. 1, pp. 179–217.

Parfit, D. (1987), *Reasons and Persons*, Clarendon Press, Oxford.

Passmore, J. (1974), *Man's Responsibility for Nature*, Gerald Duckworth & Co Ltd, London.

Peeters, W., De Smet, A., Diependaele, L., Sterckx, S. (2015), *Climate Change and Individual Responsibility*, Palgrave Macmillan, Basingstoke.

Robinson, J. (2004), *Squaring the Circle? Some Thoughts on the Idea of Sustainable Development*, Ecological Economics, Vol. 48, pp. 369–384.

Rödl, S. (2007), *Action and the First Person*. In: Rödl, S., *Self-Consciousness*, Harvard University Press, Cambridge, MA, pp. 17–63.

Rolston III, H. (1995), *Duties to Endangered Species*, Encyclopedia of Environmental Biology, Vol. 1, Academic Press, San Diego, pp. 517–528.

Rolston III, H. (1998), *Challenges in Environmental Ethics*. In: Zimmerman, M.E., Callicott, J.B., Sessions, G., Warren, K.J., Clark, J., (eds), *Environmental Philosophy: From Animal Rights to Radical Ecology*, 2nd edition, Prentice Hall, Upper Saddle River, NJ, pp. 124–144.

Rosati, C.S. (2016), *Moral Motivation, The Stanford Encyclopedia of Philosophy*, Edward N. Zalta (ed.), http://plato.stanford.edu/entries/moral-motivation/.

Rottman, J., Kelemen, D., Young, L. (2015), *Hindering Harm and Preserving Purity: How Can Moral Psychology Save the Planet?* Philosophy Compass Vol. 10, No. 2, pp 134–144.

Ryle, G. (2000), *The Concept of Mind*, Penguin Books, London.
Scarano, N. (2002), *Motivation*. In: Düwell, M., Hübenthal, C., Werner, M. (eds), *Handbuch Ethik*, Verlag J.B. Metzler, Stuttgart, pp. 432–437.
Schechtman, M. (2005), *Experience, Agency and Personal Identity*, Social Philosophy and Policy, Vol. 22, No. 2, pp. 1–24.
Scheffler, S. (2001), *Individual Responsibility in a Global Age*. In: Scheffler, S. (2001), *Boundaries and Allegiances*, Oxford University Press, Oxford.
Schlottmann, C. (2014), *Ethical Issues for Education and Climate Change*. In: Di Paola, M., Pellegrino, G. (eds), *Canned Heat, Ethics and Politics of Global Climate Change*, Routledge, London, pp. 210–224.
Schmitt, F.F. (2003), *Introduction*. In: Schmitt, F.F. (ed.), *Socialising Metaphysics*, Rowman & Littlefield Publishers, Lanham, MD.
Schmuck, P., Schultz, W. (eds) (2002), *Psychology of Sustainable Development*, Springer, New York.
Schoenmaker, D., Schramade, W. (2019), *The Principles of Sustainable Finance*, Oxford University Press, Oxford.
Schweiger Gallo, I., Gollwitzer, P.M. (2007), *Implementation Intentions: A Look Back at Fifteen Years of Progress*, Psicothema, Vol. 19, pp. 37–42.
Searle, J.R. (2005), *What Is an Institution?* Journal of Institutional Economics, Vol. 1, No. 1, pp. 1–22.
Sen, A. (1977), *Rational Fools: A Critique, of the Behavioural Foundations of Economic Theory*, Philosophy and Public Affairs, Vol. 6, No. 4, pp. 317–344.
Setiya, K. (2014), *Intention, The Stanford Encyclopedia of Philosophy*, Edward N. Zalta (ed.), http://plato.stanford.edu/entries/intention/.
Shoemaker, D. (1996), *Theoretical Persons and Practical Agents*, Philosophy and Public Affairs, Vol. 25, No. 4, pp. 318–332.
Shoemaker, D. (2015), *Personal Identity and Ethics, The Stanford Encyclopedia of Philosophy*, Edward N. Zalta (ed.), http://plato.stanford.edu/entries/identity-ethics/.
Shue, H. (1988), *Mediating Duties*, Ethics, Vol. 98, No. 4, pp. 687–704.
Shue, H. (2010), *Subsistence Emissions and Luxury Emissions*. In: Gardiner, S.M., Caney, S., Jamieson, D., Shue, H. (eds), *Climate Ethics, Essential Readings*, Oxford University Press, New York, pp. 200–214.
Singer, P. (2006), *Ethics and Climate Change: A Commentary on MacCracken, Toman and Gardiner*, Environmental Values Vol. 15, pp. 415–422.
Singer, P. (2011), *Practical Ethics*, 3rd edition, Cambridge University Press, New York.
Sinnott-Armstrong, W. (2010), *It's Not* My *Fault*. In: Gardiner, S.M., Caney, S., Jamieson, D., Shue, H. (eds), *Climate Ethics, Essential Readings*, Oxford University Press, New York, pp. 332–346.
Smith, D.W. (2013), *Phenomenology, The Stanford Encyclopedia of Philosophy*, Edward N. Zalta (ed.), http://plato.stanford.edu/entries/phenomenology/.
Smith, M. (1994), *The Moral Problem*, Blackwell, Malden, USA.
Smith, J.A., Flowers, P., Larkin, M. (2009), *Interpretative Phenomenological Analysis: Theory, Method and Research*, Sage Publications, London.
Spash, C. (2005), *Greenhouse Economics, Value and Ethics*, Routledge, London.
SRU (2002), Sachververständigen Rat für Umweltfragen, *Umweltgutachten 2002: Für eine neue Vorreiterrolle*, Metzler-Poeschel, Stuttgart.

Steg, L. (2008), *Promoting Household Energy Conservation*, Energy Policy, Vol. 36, pp. 4449–4453.

Steg, L., Vlek, C. (2009), *Encouraging Pro-Environmental Behaviour: An Integrative Review and Research Agenda*, Journal of Environmental Psychology, Vol. 29, pp. 309–317.

Steg, L., van den Berg, A., de Groot, J. (eds) (2013), *Environmental Psychology: An Introduction*, John Wiley & Sons, Chichester.

Stern, R. (2004), *Does 'Ought' Imply 'Can'?* And *Did Kant Think It Does?* Utilitas, Vol. 16, No. 1, pp. 42–61.

Strawson, G. (2004), *Free Agents*, Philosophical Topics, 32.1–2, pp. 371–403.

Stroud, S. (2010), *Is Procrastination Weakness of Will?* In: Andreou, C., White, M.D. (eds), *The Thief of Time, Philosophical Essays on Procrastination*, Oxford University Press, New York.

Stroud, S. (2014), *Weakness of Will, The Stanford Encyclopedia of Philosophy*, Edward N. Zalta (ed.), http://plato.stanford.edu/entries/weakness-will/.

Sunstein, C., Ullmann-Margalit, E. (1999), *Second-Order Decisions*, Ethics, Vol. 110, No. 1, pp. 5–31.

Thaler, R., Sunstein, C. (2008), *Nudge: Improving Decisions about Health, Wealth and Happiness*, Penguin Books, London.

Tol, R. (2016), *Comment on 'Quantifying the consensus on anthropogenic global warming in the scientific literature'*, Environmental Research Letters, Vol. 11, 048001.

Ulrich, P. (2008), *Integrative Wirtschaftsethik*, 4th edition, Haupt Verlag, Bern.

UNFCCC (1992), UN Framework Convention on Climate Change, Conference of the Parties, Earth Summit, Rio de Janeiro 1992.

UNFCCC (2015), UN Framework Convention on Climate Change, Conference of the Parties, 21st session, Paris 2015.

Unnerstall, H. (1999), *Rechte Zukünftiger Generationen*, Königshausen & Neumann, Würzburg.

Van den Hoven, M. (2006), *Common Sense Morality*, Department of Philosophy and Religious Studies, Utrecht University, Quaestiones Infinitae no. 52.

Van der Sluijs, J.P., van Est, R., Riphagen, M. (2010), *Beyond Consensus: Reflections from a Democratic Perspective on the Interaction between Climate Politics and Science*, Current Opinion in Environmental Sustainability, Vol. 2, pp. 409–415.

Van der Werff, E., Steg, L., Keizer, K. (2013), *It Is a Moral Issue: The Relationship between Environmental Self-Identity, Obligation Based Intrinsic Motivation and Pro-Environmental Behaviour*, Global Environmental Change, Vol. 23, pp. 1258–1265.

Vasileiadou, E., Heimeriks, G., Petersen, A. (2011), *Exploring the Impact of the IPCC Assessment Reports on Science*, Environmental Science and Policy, Vol. 14, pp. 1052–1061.

Vermaelen, T. (2009), *Maximizing Shareholder Value: An Ethical Responsibility?* In: Smith, N.C., Lenssen, G. (eds), *Mainstreaming Corporate Responsibility*, John Wiley & Sons, London, pp. 206–218.

Vlek, C., Steg, L. (2007), *Human Behaviour and Environmental Sustainability: Problems, Driving Forces, and Research Topics*, Journal of Social Issues, Vol. 63, No. 1, pp. 1–19.

Wallace, R.J. (2014), *Practical Reason, The Stanford Encyclopedia of Philosophy*, Edward N. Zalta (ed.), http://plato.stanford.edu/entries/practical-reason/.

WBCSD, WRI (2004), *The Greenhouse Gas Protocol: A Corporate Accounting and Reporting Standard*, revised edition.

WCED (1987), *Our Common Future*, Report of the World Commission on Environment and Development, Oxford University Press, Oxford.

Werner (2002), *Verantwortung*. In: Düwell, M., Hübenthal, Chr., Werner, M. (eds), *Handbuch Ethik*, Verlag J.B. Metzler, Stuttgart, pp. 434–445.

Wilson, G., Shpall, S. (2012), *Action, The Stanford Encyclopedia of Philosophy*, Edward N. Zalta (ed.), http://plato.stanford.edu/entries/action/.

Woods, K. (2015), *Climate Justice, Motivation and Harm*. In: Birnbacher and Thorseth (eds), *The Politics of Sustainability: Philosophical Perspectives*, Routledge, London, pp. 92–108.

## Websites

Bloomberg (2019), www.bloomberg.com/professional/blog/sustainable-debt-market-sees-record-activity-2018/, accessed 28 May 2019.

Dealogic (2019), www.dealogic.com/insight/loans-highlights-full-year-2018/, accessed on 30 May 2019.

DSM (2018), www.dsm.com/corporate/media/informationcenter-news/2018/05/15-18-dsm-links-green-house-gas-emission-reduction-to-interest-rate-in-new-euro-1-billion-revolving-credit-facility.html, accessed 30 May 2019.

EDF (2018), www.edf.fr/en/the-edf-group/dedicated-sections/journalists/all-press-releases/edf-announces-the-successful-syndication-of-an-innovative-esg-indexed-revolving-credit-facility, accessed 30 May 2019.

Elmer-DeWitt, P. (1992), http://content.time.com/time/magazine/article/0,9171, 975656-9,00.html, accessed on 28 May 2019.

ING (2017), www.ing.com/newsroom/all-news/ing-and-philips-collaborate-onsustainable-loan.htm, accessed 29 May 2019.

ING (2019), www.ingwb.com/products-services/finance-your-business/sustainable-finance/sustainability-improvement-loan, accessed 30 May 2019.

International Energy Agency (2014), *Key World Energy Statistics*, www.iea.org.

Investopedia (2019), www.investopedia.com/terms/f/freecashflow.asp, accessed on 28 May 2019.

MSF (2012), Médecins Sans Frontières, www.youtube.com/watch?v=mFq3lxdFZZ8: "Our Stories – Our People", by MSF Australia.

Philips (2017), www.philips.com/a-w/about/news/archive/standard/news/press/2017/20170419-philips-couples-sustainability-performance-to-interest-rate-of-its-new-eur-1-billion-revolving-credit-facility.html, accessed 29 May 2019.

*The Guardian* (2012), www.theguardian.com/environment/2012/jan/16/greenhouse-gases-remain-air, accessed 28 May 2019.

UDHR (1948), *Universal Declaration of Human Rights*, www.un.org/en/universal-declaration-human-rights/, accessed on 5 May 2019.

UN (1988), General Assembly Resolution 43/53, www.un.org/documents/ga/res/43/a43r-053.htm, accessed on 31 December 2018.

United Nations Development Programme (2018), Human Development Index, http://hdr.undp.org/en/composite/HDI, accessed on 31 December 2018.

Urgenda (2015), Urgenda v. The Netherlands, http://uitspraken.rechtspraak.nl/inziendocument?id=ECLI:NL:RBDHA:2015:714, accessed on 31 December 2018 (translation by the author).

Vegetarianism (2016), https://en.wikipedia.org/wiki/Vegetarianism_by_country, accessed on 14 February 2016.

World Bank (2018), https://data.worldbank.org/country/XS, accessed on 31 December 2018.

# Index